Sound Assistance

Michael Talbot Smith
BSc, CPhys, MInstP

8DP
A division of Reed Educational and Professional Publishing

A member of the Reed Elsevier plc group

OXFORD BOSTON JOHANNESBURG
MELBOURNE NEW DELHI SINGAPORE

First published 1997

British Library Cataloguing in Publication Data
A catalogue record for this book is available from the British Library

ISBN 0 2405 1439 4

Library of Congress Cataloguing in Publication Data
A catalogue record for this book is available from the Library of Congress

Printed and bound in Great Britain by
Biddles Ltd, Guildford and King's Lynn

Sound Assistance

S K I L L S E T

THE INDUSTRY TRAINING ORGANISATION FOR BROADCAST, FILM & VIDEO

What are National Vocational Qualifications (NVQs/SVQs)?
The National Vocational Qualifications/Scottish Vocational Qualifications for the broadcast, film and video industry were developed by SKILLSET to create a framework for improving the standards of vocational training throughout the industry and to establish, for the first time, a framework of professional qualifications for the industry.

Standards have been developed for every occupation in the industry, to reflect the different skills and knowledge needed at different stages in a person's career. NVQs/SVQs are work-based qualifications awarded to people who can prove that they have the skills and knowledge and are able to apply them in the workplace. Because of this emphasis on workplace assessment, NVQs/SVQs are fundamentally different from other vocational or academic awards.

The standards and qualifications will be useful for employers as a guide to levels of performance for staff and freelances. They provide, for the first time, a basis for analysing job descriptions across all sectors, which is increasingly important as we rely more heavily on a freelance workforce. The standards also provide a benchmark for the design and delivery of training of employers.

Individuals will now have proof of competence and a framework for planning further development, training and progression. This will enhance their opportunities to gain employment and to move within and outside their present area of work. Individuals will also be able to judge the relevance of different training and education programmes on offer.

How do candidates achieve an NVQ?
To achieve an NVQ candidates must prove to a qualified assessor, through a formal assessment process, that they can perform competently to the standards applicable to the qualifications sought. To do so they must produce evidence of their competence for the assessor to judge. Evidence will come primarily from the assessor's observation of the candidate's performance at work, but also from a portfolio of documentary evidence compiled by the candidate. A typical portfolio might include samples and showreels, paperwork and authenticated statements from colleagues, supervisors and employers.

Who awards NVQs?
NVQs in broadcast, film and video are awarded by the Awarding Body Partnership established by SKILLSET and Open University of Validation Services (OUVS). Candidates who gain NVQs will be awarded their certificate jointly by SKILLSET and the Open University.

What is SKILLSET?
SKILLSET is the national training organisation for broadcast, film, video and multimedia, and is funded and managed by the industry. SKILLSET's role is to:

> 'encourage the delivery of informed training provision so that the British broadcast, film, video and multimedia industry's technical, creative and economic achievements are maintained and improved.'

Contents

Introduction

The aim of this book is to give a readable and easy-to-understand account of sound operations in radio and television studios, appropriate to NVQ Level 2. It is my firm belief, shared I know by many in broadcasting, that a knowledge of what happens in a piece of equipment is essential if it is going to be used effectively.

By 'what happens' I most certainly don't mean details of the circuitry, or complexities of the hearing process and or anything of that sort. Let me explain. To take just one example, a person faced with a limiter-compressor might eventually, by simple trial and error, be able to use it to achieve certain effects. Wouldn't it be much better if that person had a mental picture of the important characteristics such as threshold and compression ratios – a visualization of a typical graph, perhaps? He or she would be much more likely to reach the wanted effects quickly. (It might even be that it was not a limiter/compressor that was needed in a particular situation but something else!) And it doesn't matter whether the limiter/compressor is an outboard analogue unit or a facility in an entirely digital console. Threshold and compression ratio are going to mean the same in either case.

Similarly, a knowledge, not necessarily very detailed, of microphone polar diagrams and how they may vary with frequency could save hours of possibly ineffectual trials.

Real theory has been excluded, although here and there I've put in small sections of explanation for the possible benefit of the more technically-minded reader but these are in clearly-marked boxes or set out at the end of a chapter so that they don't intimidate the more general user of the book.

I've tried to give practical guidance as far as possible, including some Do's and Dont's and I've emphasized the importance of Safety, stressing that safe working is as much as anything a state of mind combined with common sense.

Reference to specific types of equipment is something I've steered away from, for the obvious reason that different organizations use the products of different manufacturers. Also particular devices can soon go out of date and be replaced by improved versions.

Finally I must acknowledge, with pleasure, the great help given by Adrian Bishop-Laggett, ex Senior Sound Supervisor in BBC Television, in going through the book at the proof stage and making many useful comments and suggestions.

Michael Talbot-Smith

1

Revision of electricity – d.c.

1.1 Electricity – what is it?

An electric current consists of a stream of electrons. So, what are electrons? Well, we can't really say – and possibly no one will ever be able to describe an electron in everyday language. What we can say is that each atom of a substance consists of a core or nucleus around which electrons appear to orbit. The number of electrons depends on the substance – only one for hydrogen, eight for oxygen, twenty-nine for copper, and so on.

Each electron has a negative electrical charge and the electrons are held in their orbits, not by gravitational pull as is the case with the planets of the solar system held in their orbits by the sun, but by electrical attraction because the normal nucleus has a positive charge equal to the number of electrons.

This is a somewhat simplified picture, but it's good enough for our purposes. As to what electrons are made of, well, we don't know. Sometimes they behave as though they were solid particles; at other times they behave like little groups of waves! But for this book we can imagine that an electron is a small solid particle with a negative electrical charge. Some substances, most metals in fact, have their outer electrons rather loosely bound so that they can be detached and move on to another atom where there happens to be a vacancy. Imagine a long string of copper atoms – a length of copper wire, if you like – then given the right circumstances, which we'll come to soon, there could be a steady stream of electrons hopping from one atom to the next, and we would have an electric current.

In this chapter we deal with *direct current* (d.c.) where the current flows in one direction only, as opposed to *alternating current* (a.c.) which we'll look at in Chapter 2.

1.2 Amperes, volts and ohms

But something must cause the electrons to move in this way, and the term for that is an electrical potential difference between the ends of the wire. In other words a voltage is needed to drive a current (unit the ampere). It's helpful to think of voltage as a pressure.

It's useful to imagine a flow of water in a closed system as being equivalent to an electric current. Like almost all analogies this one doesn't have to be carried too far, but nevertheless hydraulics can give a very helpful picture as the little table tries to show.

Electricity	Hydraulic
Current	Flow of water
Ampere	Gallons/minute or litres/second
Volt	Pressure (or head of water in a hydraulic system)

To continue, a battery or other source of electricity can be compared to a pump. Further, the flow of water, or any other fluid in a system is impeded by friction, any constrictions in the pipes, and so on. The electrical equivalent is resistance.

Figure 1.1 illustrates the electricity/fluid analogy.

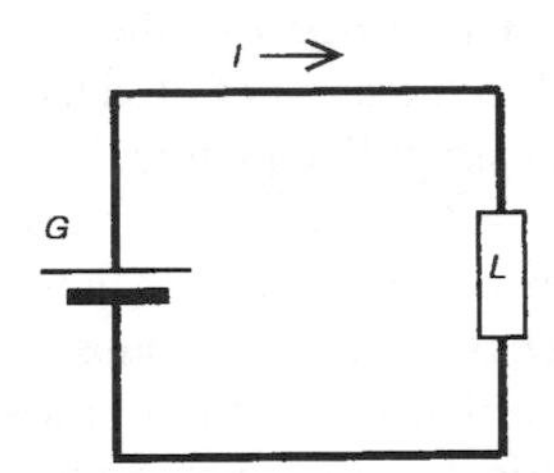

Figure 1.1 (a) An electrical circuit

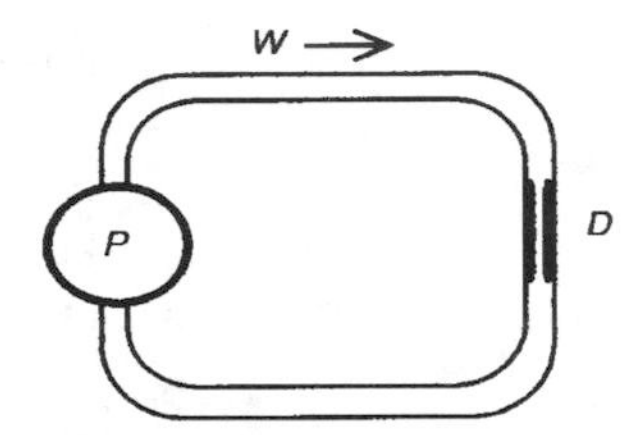

Figure 1.1 (b) A hydraulic circuit

In (a) G is an electrical generator (for example a battery or a dynamo). Its equivalent in (b) is a pump, P. I represents a current (amperes), W is water flow, while L is a load or any other device presenting resistance (a motor or a heater, say) and is equivalent to (in this case) D, a device which restricts the water flow. It is shown here as a constriction in the pipe.

The unit of resistance is the ohm, symbol Ω (the Greek letter omega) and amperes (A), volts (V) are related to it by the formula:

OHM'S LAW
$I = V/R$
$V = IR$
$R = V/I$

amperes = volts/ohms

This important formula, known as Ohm's Law, is usually written as $I = V/R$ where I is the general symbol for current, and V and R represent voltage and resistance respectively.

The formula may be written in two other ways: $V = I \times R$ and $R = V/I$.

Example: A 2 V battery is connected to a small lamp. A meter shows that the current is 0.5 A. What is the resistance, under these conditions, of the lamp?

Answer: Using the formula $R = V/I$, if V = 2 V and I is 0.5 A then R is 2/0.5 = 4 A.

(Note that we used the phrase 'under these conditions'. The point is that the electrical resistance of most substances increases with temperature, so if the small lamp is glowing brightly its resistance will be much higher than when there is insufficient current to make it light up. The cold resistance of the lamp in the question might be no more than 1–2Ω.)

1.3 Power and watts

Power is the rate at which work is done. If an electrical device is consuming a lot of power then it ought to mean that it is doing a lot of useful work. A high-power electrical motor, for instance, could be used in a crane to lift heavy loads. The unit of power is the *watt*, symbol W. As far as we're concerned here we can say that basically the number of watts is found by multiplying amperes by volts:

$$W = I \times V$$

In the example above the lamp has 2 V across it and the current is 0.5 A. The electrical power it consumed is then:

$$0.5 \times 2 = 1 \text{ W}$$

But don't be misled into thinking that it was emitting 1 W of light power. It would be producing a total of 1 W of *energy* of which perhaps less than about one-tenth would be in the form of visible light!

From $W = I \times V$, and remembering Ohm's Law, we can do some substitution and find different formulae for power:

Since $I = V/R$ we can replace the I in $W = I \times V$ with V/R, making:

$$W = V/R$$

and, since $V = I \times R$ we can replace the V in $W = I \times V$ and thus arrive at:

$$W = I \times R$$

POWER
$P = IV$
$P = I^2R$
$P = V^2/R$

The reader is advised to learn the three forms of the formula for power as well as the three forms of Ohm's Law. (Actually, you don't really need to memorize them all – if you know the basic forms then it's easy to arrive at the other versions! However it's very useful to be able to recognize them.)

A word or two of caution. If we're dealing with alternating current the formulae we've given may not be correct for all a.c. conditions although they are correct in d.c. circuits.

1.4 Ideas about magnetism

Similar poles repel
Dissimilar poles attract

We're all familiar with magnets and the fact that they can attract certain metals to them.

It's standard to identify the two ends of a magnet – the poles – with the letters N and S, short for North and South. This is because if a bar-type magnet is suspended at its centre by a long and untwisted piece of thread the N pole will swing to and fro and finally come to rest pointing towards the North. The Earth is itself a magnet, although the reasons for this are still rather obscure. Now it is found that similar poles, such as two N poles *repel* each other but unlike poles, an N and a S, are *attracted* to each other. This leads to the slightly confusing fact that the Earth's north magnetic pole (which is near to but not at the true North Pole) is, magnetically, a South pole!

Some of the most useful magnets in practice are those which are curved round so that the N and S poles are, as it were, facing in the same direction. Such magnets are generally called *horseshoe* magnets, although the similarity with an actual horseshoe is sometimes slightly fanciful.

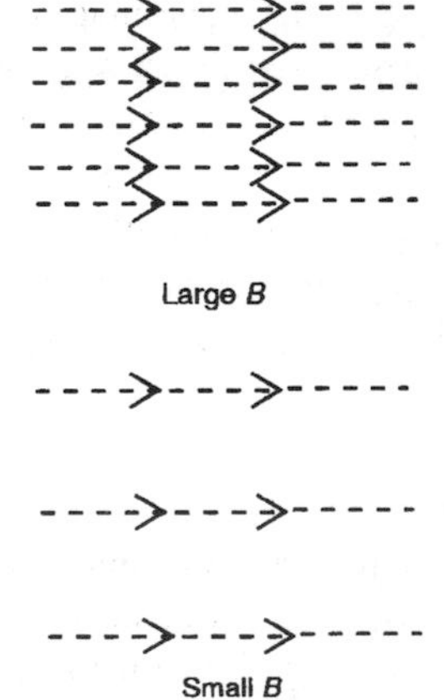

The region in which the effects of a magnet can be detected is called the *magnetic field* and this is represented in diagrams by a number of lines, going from the N pole of a magnet to the S pole.

The closeness of the lines show the *magnetic flux*, that is, the strength of the magnetic field.

The symbol for magnetic flux is ϕ and the unit is the *weber*, Wb. We give a definition at the end of the chapter. There then follows the idea of flux density, which can be thought of as the concentration of flux. The fairly obvious unit of flux density, symbol $\boldsymbol{B}$, is the *weber/meter*^{2}.

The weber is a large unit and small fractions of a weber are normal in audio. For example, it's useful to be able to specify the intensity of magnetization of a recording tape. Here, the flux across the tape is a handy concept as this gives an indication of the signal which will be generated in a replay head. The unit in this case is based on *flux/metre*, but webers/metre is too large to be useful. Instead *nanowebers/metre* or *nWb/m* is appropriate, the prefix *nano* meaning 10^{-9}. A typical recorded signal will have a flux in the region of 300 nWb/m.

Webers and its sub-units are not necessarily something that the reader must remember. We mention them because sooner or later you will meet them, and just to have met the terms before makes the encounter less alarming!

1.5 Magnetic fields and electric currents

Magnets provide a permanent and essentially constant magnetic field. Sometimes this is exactly what is wanted, as we shall see in things like loudspeakers and some microphones. At other times we may want a field that is variable and under a degree of control. Here it's fortunate that when a current flows in a wire, or any other conductor, a magnetic field is generated around the conductor. The field round a single conductor will probably be very weak, but if the conductor is wound into a coil the whole assembly begins to behave very much like an actual magnet, especially if the coil is wound round a suitable magnetizable substance such as iron. Here we must distinguish between different types of magnetizable substance. We shall talk about magnetically *soft* materials, which when placed inside a coil carrying a current, are magnetized *only when there is a current flowing*. As soon as the current is switched off the substance loses all, or almost all, of its magnetization.

Then there are magnetically *hard* materials, and these retain their magnetism when the current is switched off, or the material is removed from the coil. There are uses for both types. Loudspeakers use *permanent* magnets, and these can be made using a suitable coil and a magnetically hard substance. The *recording head* of a tape machine consists basically of a coil round a magnetically soft substance so that the field depends on the current flowing in the coil and ceases when there is no current.

1.6 Currents in coils

An electromotive force (e.m.f.) is measured in volts and the distinction may seem rather pedantic. However, it is customary, and can reduce confusion, if we use the term e.m.f. to mean the voltage produced by a generating device before any current is taken from it. (A fuller definition of e.m.f. is given at the end of this chapter.)

This property is made use of in some kinds of microphone, where sound waves cause a very thin diaphragm to move. Attached to the diaphragm is a small coil in the field of a permanent magnet. The e.m.f. induced in the coil is proportional to the pressure of the sound waves striking the diaphragm.

A further effect is that if we have a conductor in a magnetic field and cause a current to flow in it then there is a force upon the conductor, tending to make it move. The force is proportional to the flux density, the current and the length of the conductor in the field. Most loudspeakers make use of this effect.

The great majority of electrical measuring meters also use the effect. The magnet is of the 'horseshoe' type and the poles are specially shaped with a cylindrical core of magnetic material. This concentrates the flux and also helps to make the flux density reasonably constant over the region in which the coil moves. In this way the deflection of the pointer is proportional to the current in the coil. There are spiral hair-springs (there may be more than one) which act against the force on the coil when there is a current in it. If this is not done then the pointer will not return to its rest position when the current ceases.

1.7 Protecting equipment – fuses

With the exception of a few items such as pocket torches almost all electrical devices are 'fused' – that is they contain somewhere a small component consisting of a length, perhaps a centimetre or two, of special wire designed to melt when the current flowing in it exceeds a certain value. The current supply to the device flows through the fuse so that when, for whatever reason, the current exceeds safe limits the fuse melts (we say it 'blows') and the current is then cut off. It is vitally important that the correct value fuses are fitted, especially when the equipment is powered from the mains. Let's take an example: a domestic television set might be 'rated' at 300 W. This means that, with the electricity supply voltage being about 230 V, the current taken by the set will be a little more than 1 ampere. (Actually the mains supply is a.c. but that doesn't affect the calculation that we're doing here.) Suppose the set developed an internal fault causing a short circuit – meaning that there is insufficient resistance to limit the current significantly. The excessive current could easily cause over-heating leading to a disastrous fire.

However, with a fuse rated at, say 3 A, the current cannot be greater than this before the fuse melts and cuts off the supply.

With almost all domestic equipment the fuse is inside the mains plug. Unfortunately plugs tend to be supplied with a 13 A fuse already fitted in them and this is much too large for very many items. It's always important to check that the correct fuse is fitted. For domestic purposes the most common fuse ratings are 1, 3, 5 and 13 A. It's easy to check: divide the power taken by the equipment – and this is always stated somewhere – by the mains voltage to find the current. Then, normally, the next size fuse up will be right. (There are some items that take a large initial current – a 'surge' – and this must be allowed for. Usually the manufacturer's state the correct fuse rating.)As examples, a 300 W television set should probably have a 3 A fuse; a 100 W lamp should have a 1 A fuse, and so on. *But*, and it's a big but, remember that fuses protect the equipment and its cabling, not people directly! They are safety devices in that they can reduce the risk of a fire, which of course is therefore a safeguard for people, but don't imagine that if you grasped the wires from a mains plug the fuse would blow and protect you. You'd probably be dead long before even a 1 A fuse blew! We'll deal with other protecting devices in the next chapter.

1.8 Batteries

Actually we generally use the term 'battery' incorrectly!

A *battery* is strictly a number of individual *cells* but we have become so used to going into a shop to ask for 'a such-and-such size battery' for, say, a camera that the habit is not easy to lose. What we should do is to ask for 'an AA cell', or whatever, but one fears that there may often be a blank look from the other side of the counter! Here we'll try to be correct and refer to *cells* when we mean single units.

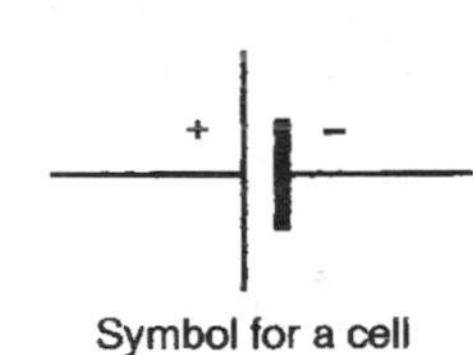

Symbol for a cell

There is a very great range of types of cell but here we will only deal with the ones that the reader is most likely to encounter. And broadly they can be divided into two kinds: first there are *primary cells*. In these the cell is assembled and is immediately capable of supplying electricity as a result of chemical reactions. The ability to generate a current continues until one or other of the chemical constituents is used up. *Secondary cells*, on the other hand, need to be *charged* by passing a current through them. Once charged they can supply a current until they are *discharged*, but they can then be recharged. Normally the charging/discharging

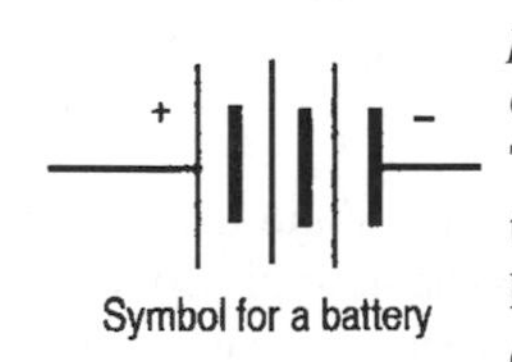

Symbol for a battery

process can be carried out hundreds or even thousands of times.

There are a few very important characteristics of a cell. The first and most obvious one is the terminal voltage. We should use the term *e.m.f. – electromotive force* – meaning the voltage at the terminals when no current is flowing. The next most important characteristic is the *internal resistance*. When a cell is supplying a current it must be remembered that the current flows through the cell as well as the external circuit. And there is always some resistance inside the cell as well as outside. The effect of the internal resistance is two-fold: it reduces the terminal voltage when a current is taken from the cell and it also determines the maximum current which can be supplied.

Let us take an example: a cell has an e.m.f. of 1.5 V and it has an internal resistance of 2 Ω.

1. Suppose the cell is connected to a resistance of 2.5 Ω. What current will flow?

The possibly immediate, but wrong, answer is to apply Ohm's Law by saying:

$$I = V/R = 1.5/2.5 = 0.6 \text{ A}$$

This ignores the internal resistance of 2 Ω. The true answer is based on the fact that the total resistance is 2 + 2.5 = 4.5 Ω. The correct answer for the current is then:

$$I = 1.5/4.5 = 0.33 \text{ A}$$

2. The maximum current that the cell can deliver is found by taking the external resistance to be zero, so the total resistance in this case is 2 Ω:

$$I = 1.5/2 = 0.75 \text{ A}$$

We can now give a brief outline of the types of cell commonly met with:

Standard primary cells (e.g. sizes 'AAA', 'AA', 'C' and 'D'): The e.m.f. is 1.5 V and the internal resistance is typically around 1 ohm. 'PP9' or equivalents are true batteries, being made up of six cells. The e.m.f. is thus 6 V, although the internal resistance may be more than 6 Ω.

Most primary cells have e.m.f.s in the range 1 to 2 V

Alkaline cells: the e.m.f. is again around 1.5 V but the internal resistance is lower than in standard cells, being typically 0.7 Ω or less. Alkaline cells are much more expensive than standard ones but have a much greater capacity – that is they can deliver a particular current for far longer than a standard cell, perhaps as much as twenty times. It is usually found that they are well worth the extra cost.

Lead–acid cells: these are secondary cells and are thus rechargable. The best-known is probably the car battery, made up of six cells, each with an e.m.f. of 2 V, making 12 V total. Small batteries of 2, 4 or 6 cells are widely used for powering things like portable television cameras and video recorders. Lead–acid cells have a very low internal resistance – usually a small fraction of an ohm. This means that if they are accidentally short-circuited – for example their terminals are connected by a simple piece of wire – then a very large current can flow, resulting possibly in the wire melting and a consequent risk of fire or other damage. This type of cell should not be left uncharged for any length of time as permanent damage may occur.

Nickel–cadmium (ni-cad) cells: for most purposes these are rechargeable substitutes for standard or alkaline cells. Their e.m.f. is lower – about 1.2 V compared with 1.5 V – but their internal resistance is extremely low and this can to a large extent compensate for the lower e.m.f., in that when delivering a moderate current the terminal voltage does not fall as much. Because of this low internal resistance short circuits have to be avoided. Charging must be carried out strictly following manufacturers' instructions. This generally means charging slowly over a longish time – 16 hours is often quoted – but there are clever 'fast-charge' devices available.

Disposal: When finished with ALL batteries should be put safely into an appropriate container. Some contain toxic substances; many, if over-heated, can explode and the explosion may scatter dangerous materials.

Always treat discarded cells and batteries with caution.

Some prefixes

Prefix	Symbol	Meaning	Written
mega	M	x 1 million	x 10^6
kilo	k	x 1 thousand	x 10^3
milli	m	x 1 thousandth	x 10^{-3}
micro	µ	x 1 millionth	x 10^{-6}

There are other prefixes for larger or small quantities but the ones here are enough to be going on with. We'll deal with others if and when we need to.

Definitions

Magnetic flux

The weber, is defined as the flux that, if reduced to zero in one second, when linked with a coil of one turn, will induce an e.m.f. of one volt.

e.m.f.s in coils

The e.m.f. generated in a conductor of length l, in a magnetic field of flux density $\boldsymbol{B}$, when the velocity v of the conductor relative to the field, measured at right angles to the direction of the lines of flux, is given by $e = \boldsymbol{B}lv$.

Flux density and newtons

For a conductor of length l carrying a current of I amps in a magnetic field with a flux density $\boldsymbol{B}$ the force on the conductor at right angles to the direction of the flux is:

$$F = \boldsymbol{B}Il \text{ newtons}$$

The weight of an apple is roughly one newton

(The newton is the unit of force and its strict definition is that it is the force which would give a mass of 1 kg an acceleration of 1 metre per second per second. An easier concept is that an average size apple held in the hand has a downwards pull, caused by gravity, of about one newton.)

2

Revision of electricity – a.c.

2.1 The 'shape' of an alternating current

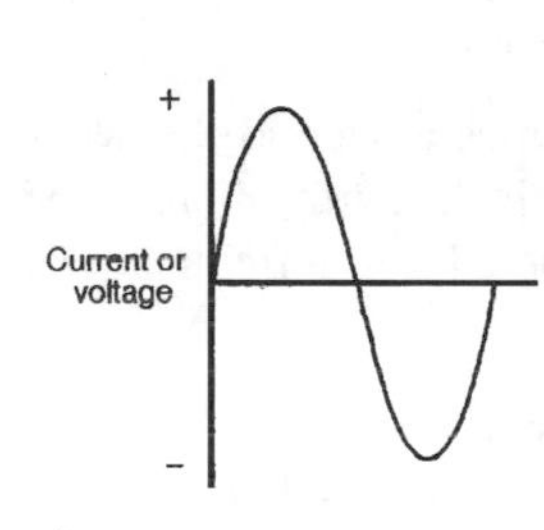

If we say that something like an electric current is alternating, this simply means that the current first goes in one direction and then the other. However, in electricity and many other applications, the term has a somewhat specialized meaning.

Of course the voltage or current represented in the diagram is first going in one direction and then the other, but it is doing it smoothly. This type of waveform is called a sine wave, because it can be obtained by plotting the trigonometric quantity called a sine against, in this case, time. A very brief explanation of the term 'sine' is given at the end of this chapter in Section 2.13.

The reason why sines are significant arises from the basic method of producing an alternating current. We saw in Chapter 1 that an e.m.f. is generated when a conductor moves in a magnetic field. If the conductor is in the shape of a coil and this rotates (Figure 2.1) then the resulting e.m.f. has the shape of a sine wave.

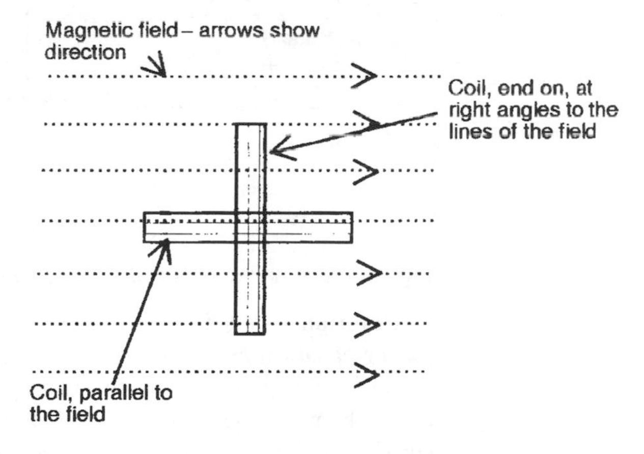

Figure 2.1 A coil rotating in a magnetic field

When the coil is vertical the conductor is moving along the field and no e.m.f. is produced. The maximum e.m.f. occurs when it is horizontal. A little imagination shows that the e.m.f. at the terminals of the coil is going to have the shape of a sine wave.

2.2 Important quantities

With d.c. it is normally quite enough to state the voltage (or maybe the current) to give adequate information. With a.c. it is clearly not quite so simple. The important quantities – measurements or features, if you like – are:

1. The frequency – how many times a second the wave shape is repeated. The basic unit is the hertz, abbreviated to Hz. The electrical supply in the UK has a frequency of 50 Hz.

   For much higher frequencies we use:
   kilohertz (kHz) = thousand Hz
   Megahertz (MHz) = million Hz
   Gigahertz (GHz) = thousand million Hz.

2. The amplitude or peak, not to be confused with the peak-to-peak or double amplitude voltage or current. (See Figure 2.2.)

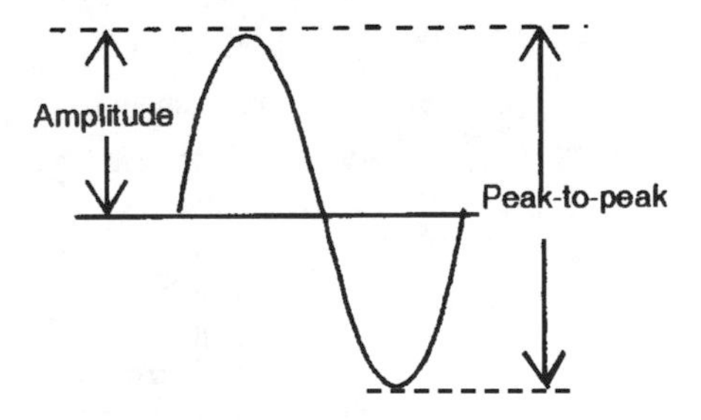

Figure 2.2 Amplitude and peak-to-peak

Unfortunately the amplitude tells us very little about the power capabilities of an alternating current. If we want the current to do something useful, like operate a light or a motor, all we know at the moment is what the maximum voltage or current are, but we also know that a fraction of a second latter these are going to be zero! This brings us to the third important quantity.

3. The *root-mean-square (r.m.s.)* values. This sounds complicated.

What it means really is this: The r.m.s. value of an a.c. current or voltage is that which would have the same heating effect as a d.c. current or voltage.

For example, a d.c. supply of 12 V can, when connected to a lamp bulb, produce a certain brightness. An a.c. supply, connected to the same bulb, would produce the same brightness if its r.m.s. voltage were 12 V. Its *amplitude*, or peak, would be much more than 12 V – nearer to 18 volts in fact, but of course much of the time the voltage would be very much less than 12 V.

For sine waves the relationship between the r.m.s. voltage and the peak voltage (amplitude) is:

$$V_{r.m.s} = 1/\sqrt{2}V_{peak}$$

The mains supply in the UK is quoted as 230 V. This is the r.m.s. value and a d.c. supply of 230 V would produce the same power in a heater. The peak value of the mains is therefore:

$$\sqrt{2} \times 230 \text{ V}$$

Since $\sqrt{2}$ is 1.414, the peak value of the mains voltage is about 230 x 1.414 = 325 V.

The term 'root mean square' arises in the following way: a straightforward average of a sine waveform would be zero, as the positive part equals the negative part. But by squaring the individual values they all become positive. Taking the average, or mean of these squared values, and then compensating for the squaring process by taking the square root, gives us a sensible value. It can be shown, although we shall not do so here, that the process also gives the d.c. value which is equivalent in power producing terms.

The reader with curiosity and a scientific calculator can try the effect of finding the sine of 0°, squaring it, then finding the sine of 10°, squaring it and so on at 10° intervals to 180°, then adding all these squared numbers, calculating their average, and finally finding the square root of the average. It will be reasonably close to 1.414.

2.3 Phase angles

As we have said, the complete sine wave in Figure 2.1 represents a rotation through 360° of a coil. It's therefore reasonable to see the horizontal axis (the *x*-axis) as being marked in degrees, as in

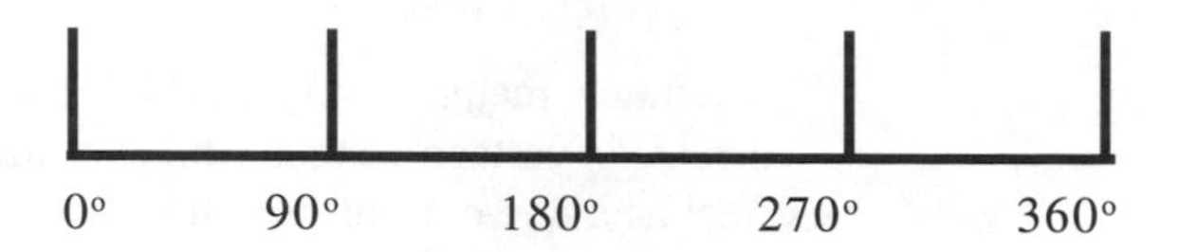

Figure 2.3 Using degrees as a scale

Figure 2.3. (There are, of course, other equally valid scales. Time is a commonly-used one.)

This gives us a convenient way of expressing the relationship in, for example, time, between two different sine waves – as a *phase angle.* The margin diagram illustrates this.

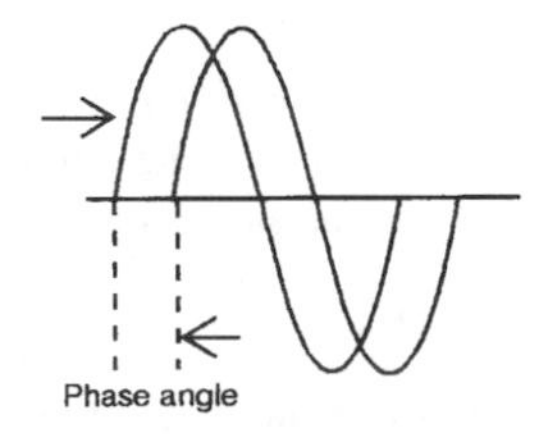

In our little diagram the *phase difference* between the waveforms is 90°. Of course, if we know the time for one cycle – the *period* of the waveform – it's quite easy to convert a phase angle into a time difference. For instance, if the frequency of the waveforms in Figure 2.3 is 50 Hz then the period is 1/50 s or 20 milliseconds, abbreviated to 20 ms. The 90° phase difference in this case is equivalent to a time difference of 5 ms. (Note that the two waveforms can have quite different amplitudes and we can still talk about their phase difference. If their frequencies are not the same then phase angles become pretty meaningless.)

2.4 Three-phase supplies

Large-scale generation and distribution of electricity is normally done on a three-phase system. If we think initially of three pairs of cables then each pair carries a 50 Hz current but the pairs differ from each other in having a 120° phase difference.

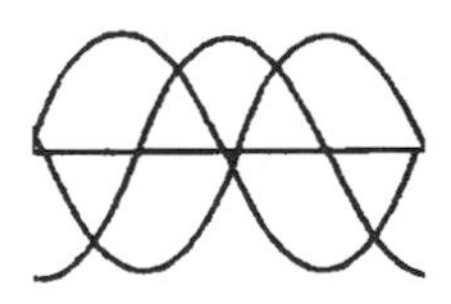

It should be reasonably clear, by looking at the margin diagram, that the *sum* at any instant of the three currents (or voltages) will be zero, but this will only be true if the three phase voltages or currents are exactly equal.

When that is the case they can be suitably joined together to have one common wire and the 'return current' in it is zero. In practice this is never the case, but by careful planning the return current can be small. In some overhead distribution systems, especially in rural areas, there are only four cables, three moderately thick for the three phases and one rather thinner for

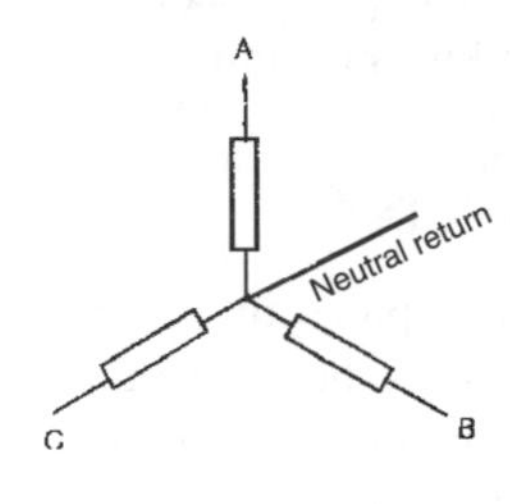

the return. A typical way of bringing the three phases together is by a *star* connection, shown on the left.

Now if each phase voltage is 230 V then there is a higher voltage between A and B, B and C, and C and A.

It is √3 x 230, or 1.73 x 230, which is approximately 400 V.

There are many advantages in having a three-phase supply going into a factory: besides the standard 230 V supply there is also 400 V for machines running at higher power; also three phases supplied in the correct way to a suitable motor can give a smoother torque and motors can be smaller. A television studio complex may have different studios on different phases to help balance the loads – i.e. make the return current as small as possible.

Don't mix phases!

But, be warned! It can be very dangerous to have equipment in one studio supplied with power from another. If we have two items, one powered from one phase and the second powered from a different phase there can be 400 V between them!

2.5 Mains plugs

Correct wiring of these is vital:

The *live* (L) terminal, which is connected to its brass 'pin' via a fuse, must have the *brown* wire attached to it, the *blue* wire goes to the neutral (N) connection and the *earth* wire (green/yellow) goes to the connector nearest the apex of the plug and marked E. (Older cabling is still found in which the live wire is red, the neutral is black and the earth is green.)

The bared cables must be firmly screwed into the connections and the cable insulation must only be removed as far as is necessary – there shouldn't be any copper visible! It is often recommended that there should be a little slack in the earth and neutral wires but *not* in the live, so that if the cable is pulled excessively it is the live wire which comes away first. However cable grips must always be fastened down securely.

It is vital that all plugs are checked regularly, especially if they are attached to equipment which is portable.

2.6 Safety devices

In Chapter 1 we dealt with fuses and we made the point that they protect equipment, initially. People are really only protected because fuses reduce the risk of electrical fires.

While we are on the subject of mains electricity we can

mention devices that can, up to a point, safeguard people from electrical shock. Broadly there are two kinds of unit:

1. residual current devices (RCDs),
2. earth leakage current devices (ELCDs)

If mains electricity is flowing into a piece of apparatus, whether it is an electric kettle in the home, a hedge trimmer in the garden or an item of recording equipment in a studio then the current going into the apparatus from the *live* side of the mains should exactly equal that returning to the *neutral* side. If, however, there is a fault which could be dangerous to a person, such as the casing of the apparatus becoming live, then there will be a flow of current to earth. RCDs and ELCDs detect this imbalance. RCDs compare the current going in from the live side with that returning to the neutral and if they are not equal a 'trip' operates and cuts the supply. ELCDs operate by detecting a flow of current to earth. It is a requirement in the UK that either kind of unit operates when the current imbalance (or earth current) exceeds 30 mA and, what is more, does so within one-thirtieth of a second. Thus, suppose the metal casing had become live and a person provided some sort of an earth connection by touching it, the safety device would operate almost instantly and before the person had received a dangerous shock. (No protection is given if someone touches the live and neutral sides of the supply!)

It is becoming increasingly common for mains supply units feeding a technical area to be equipped with RCDs and this is also becoming true in domestic situations. Small, portable, RCDs are widely available and these should *always* be used with mains equipment out-of-doors.

2.7 Transformers

A big advantage, probably the biggest, of a.c. over d.c. is that voltages and currents can be easily changed with very little loss of power by *transformers*.

In its simplest form a transformer consists of two coils of wire wound on an core of suitable iron.

The two coils have different numbers of turns on them. The coil into which the power is fed is called the *primary*, the one from which power is taken is called the *secondary*, and the ratio:

Number of turns in the primary/number of turns in the secondary is called the *turns ratio*. It is usually denoted by n.

The turns ratio determines the amount by which the voltage is *stepped up* or *stepped down*.

Suppose we wish to use the a.c. mains to operate a 12 V lamp. The transformer we would use would have a step-down ratio of:

$$n = 230/12 = 19.1$$

i.e. there would be 19.1 times as many turns on the primary as on the secondary.

Exactly how many turns on each coil would be used depends on many things. With large numbers of turns there will be greater efficiency. On the other hand, the more wire used the greater the losses resulting from heating of the wire unless is it very thick – and then size, weight and cost go up!

In the electrical distribution system the size and weight are not primary problems and the cost of a big transformer using large diameter wire will probably be easily justified by the greater efficiency.

A transformer used to step down, say, 132 000 V on the grid system may be handling powers of thousands of kilowatts. A 1% loss represents a few tens of thousands of watts!

A small transformer used in audio work could have appreciable power losses without anyone worrying – what may well be much more important there is the ability to handle a wide range of frequencies.

It can generally be assumed that the losses in a good transformer are small, so that the power that comes out is almost the same as the power that goes in:

$$\text{Power}_{in} = \text{Power}_{out}$$

or

$$V_{in} \times I_{in} = V_{out} \times I_{out}$$

Suppose we have a transformer with a turns ratio of 20:1 the voltage at the primary is 50 V with a current of 2 A. The power going in is then 100 W and we will take it that the secondary is delivering 100 W also (not quite true but with a reasonably good transformer, even a small one, the efficiency could well be 98%, so that, in this example, only 2 W would be lost as heat). The voltage is stepped down 20 times, from 50 to 2.5.

If the voltage at the secondary is 2.5 V and the power is

100 WV, then the secondary current must be 40 A: we have stepped the current *up* by the same proportion that the voltage was stepped down.

Transformers can thus be:

- voltage step-up
- voltage step-down
- current step-up
- current step-down

(Not all at the same time, of course!)

The transformers used for arc welding are good examples of current step-up. 230 V mains goes in and a very low voltage but very high current comes out.

2.8 Meters for a.c.

In Chapter 1 we showed the principle of a moving coil meter. These meters work perfectly well with d.c., but on a.c. the pointer can do nothing except vibrate rapidly about its zero, possibly so rapidly that it's impossible to detect any movement at all.

However, moving coil meters can be used for a.c. but with a modification. This is an arrangement of rectifiers – components which allow a current to flow through them in one direction only.

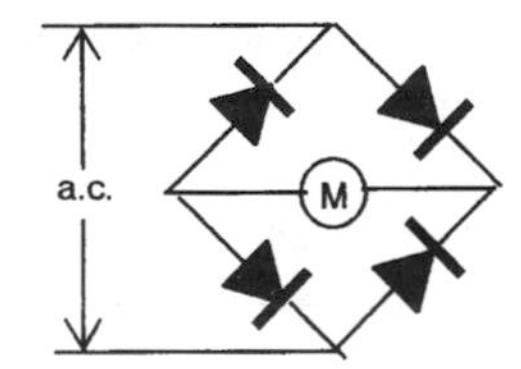

The rectifiers are arranged as shown in the margin diagram. The arrowhead in the symbol shows the direction that conventional current will flow through the rectifier (not the electrons – they go in the opposite direction!).

The reader should satisfy him/herself that a current will go in only one direction through the meter. (Does it go from left to right or right to left?)

There is a slight problem with a.c. meters – the characteristics of rectifiers are not linear: the relationship between voltage across them and current though them is not quite a straight line and this is particularly true at low voltages. As a result the meter scale tends to be slightly cramped at the low end of the range. This is seen in meters which can be used for both d.c. and a.c. measurements – there are different scales for the two.

Most of this chapter has so far been dealing with topics that are mainly related to large, or relatively large power systems, supply mains voltage and upwards. Now we turn to components that, while they can have a place in heavy current engineering, are mostly found in electronic circuits.

2.9 Capacitors

A capacitor is a device which can store electricity, but, unlike a re-chargeable battery, in small or even very small quantities only. The diagram shows the most basic type of capacitor – two conducting plates, insulated from each other.

If electrons (negative charges) are put on to the right-hand plate, by, for example connecting that plate to the negative terminal of some kind of d.c. supply, then they exert a repulsive force on electrons on the left-hand plate and if that plate is connected to the positive terminal of the supply these electrons leave the plate and there is a momentary current flow. It's only a brief current because once the repelled electrons have left their plate everything stops and the only way to get the current to flow again would be to increase the voltage of the supply device so that, for an instant, there would again be a current.

We say that the capacitor now has a charge on it. This charge, *Q coulombs*, the applied voltage, *V*, and the *capacitance*, *C*, are related by:

$$Q = CV$$

A coulomb is a unit of quantity of electricity and is equal to a current of 1 A flowing for 1 second.

The unit of capacitance is the *farad*, which commemorates, albeit incompletely, Michael Faraday, sometimes called 'The Father of Electricity'.

A practical definition of the farad is that a capacitor has a value of 1 F if the potential across it is 1 V when there is a charge of 1 C on it. (This, it will be noticed, is simply using the formula above to define the farad.)

Unfortunately 1 F is inconveniently large. More practical values are:

- microfarad (μF) which is one millionth of a farad, or even
- picofarad (pF), which is one millionth of a microfarad (10^{-12} F).

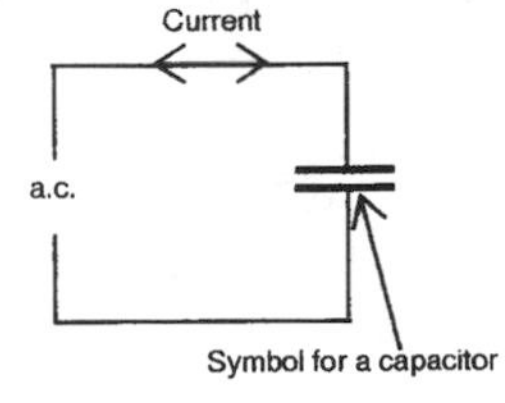

The applications of capacitors in electronics are far too numerous to go into here, but we can outline a very important aspect – the behaviour of capacitors in a.c. circuits.

To begin with, if the polarity of the supply is changing constantly, as it is in an a.c. circuit, then a capacitor in series with that supply will be charging and discharging repeatedly, so that a current will flow. It's quite common, but incorrect, to talk of an a.c. current 'flowing through' a capacitor. It doesn't really – charge flows on to one plate and away from the other, but it's as if a current actually did flow through the capacitor.

Now it ought to be fairly obvious that that larger the capacitor (i.e. the more farads, or, probably, microfarads) then the larger will be the current that will flow into and out of it.

It ought also to be unsurprising that if the frequency of the a.c. is higher then the current will also be higher – the more rapid the rate of alternation then the greater the flow of electrons into and out of the capacitor. This leads us to the idea of reactance.

Reactance is a little like resistance but there are a few major differences. We use the symbol X for reactance, often with a subscript, but more of that later.

Now, as we have seen, resistance is equal to V/I. It is the ratio of voltage to current.

Reactance is, in formula terms, exactly similar:

$$X = V/I$$

and the unit of reactance is still the ohm.

The differences between resistance and reactance are:

1. Whereas resistance implies that there is a heating effect (power losses, if you like, although these might be intentional) there is no power loss, or any heating effect, in an ideal reactance device such as a pure capacitor. There may be some power losses in a practical case but the hope is that these are very small and often can be ignored.

2. The reactance varies with the frequency.

From now on we shall use the notation X_c for *capacitive reactance*.

We have said that the current through a capacitor increases with the value of the capacitor, and it also increases with the frequency. If the current (by implication for the same voltage) increases then the reactance must decrease, and an expression for capacitive reactance is:

$$X_c = 1/2\pi fC \text{ ohms}$$

This is quite an important formula.

2.10 Inductors

Basically an inductor is a coil of wire, often wound round a core of magnetic material.

A direct current in an inductor simply creates a steady magnetic field and there is then no further change. However, with a.c. the field is constantly changing and in doing so it generates within the coil an e.m.f.

What is very important to realize is that this e.m.f. is in such a direction as to oppose the voltage across the inductor.

This means we have a reactance – in this case inductive reactance and we'll use the notation X_L to indicate it. (L is a standard letter for inductance – goodness knows why!)

Let's apply common sense, as we did with capacitors, to try to see what factors are going to affect the reactance of an inductor.

If we increase the number of turns on a coil then we're going to increase any magnetic effects, and make large the generation of the opposing e.m.f. So X_L will increase with L.

If the frequency increases the rate of change of the magnetic flux will also increase and again there will be an increase in reactance. In other words, X_L will increase with f. The complete expression is:

$$X_L = 2\pi fL$$

We haven't said yet what the unit of L is. It has been given the name of an American physicist, (Joseph) Henry. An inductance has a value of 1 H if the e.m.f. across it is 1 V when the current in it is changing at the rate of 1 A per second.

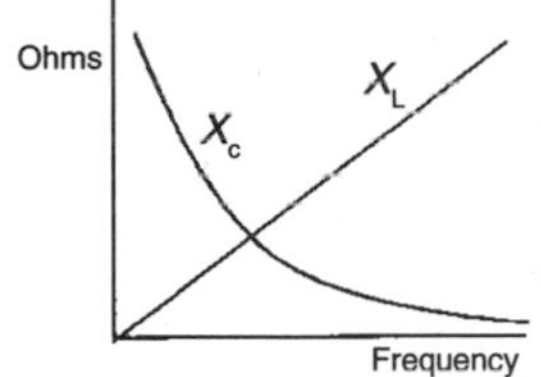

Notice that for a capacitor reactance decreases with frequency but it increases for an inductor. A convenient way of showing this is with a graph, as in the small diagram.

The reader will no doubt realize that in capacitors and inductors we have components which can be a very useful basis for all kinds of frequency-selective devices – filters for reducing the effects of unwanted background noises, tone controls and means of boosting weak signals in a particular frequency range. (We should add that capacitors are generally used much more, being smaller, lighter and cheaper, and can, with the use of a bit of ingenious circuitry, do most of these jobs.)

2.11 Trigonometry

Trigonometry, essentially the mathematics of triangles, makes use, among other things, of the ratios of the sides of right-angled triangles.

The three most commonly used trigonometric ratios are:

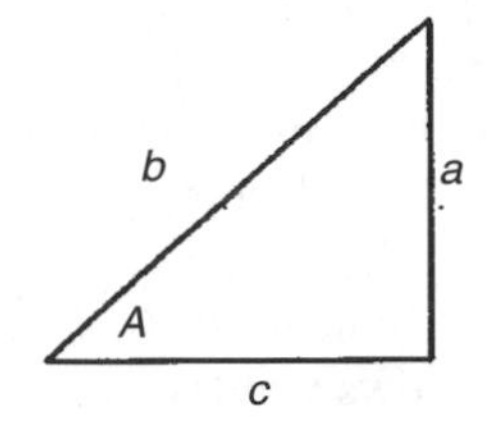

1. The *sine*. In the small diagram the sine of the angle A is the ratio of the opposite, a, to the hypotenuse, b. 'Sine' is usually abbreviated to 'sin'.

$$\sin A = \text{opposite/hypotenuse} = a/b$$

2. The cosine, abbreviated to 'cos'.

$$\cos A = c/b$$

3. The tangent, abbreviated to 'tan'.

$$\tan A = a/c$$

3

Outlines of electronic devices

It would be impossible in this book to do more than give the barest outline of electronics. It would also be inappropriate. This chapter is therefore going to be little more than a glossary of some of the more important terms used in electronics.

3.1 Semiconductors and thermionic devices

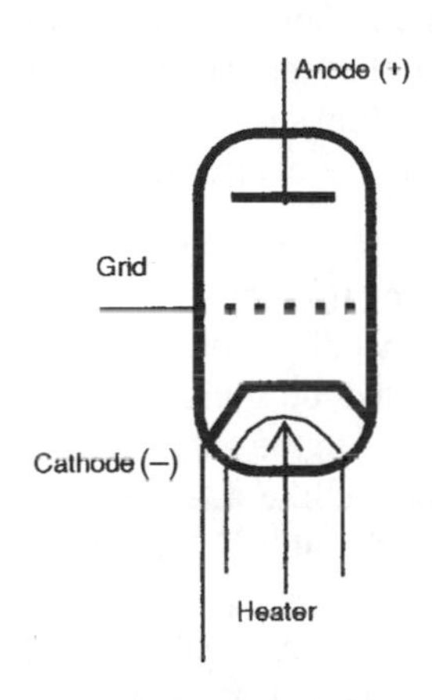

Symbol for a triode valve

Thermionic devices: 'Valves' (or 'tubes' in the USA) came first. Inside a highly-evacuated glass container an electrode called the *cathode* is heated and, because of this, it emits electrons. These are attracted towards a second electrode called the anode which is at a relatively high potential. This could be anything from about 10 V up to many kilovolts. The arrangement rectifies because electrons cannot leave the cool anode. Between the cathode and the anode are one or more open-mesh electrodes, called *grids*, which can control the electron flow from cathode to anode.

Amplification is possible by applying a small varying voltage to a grid. There may be one or more grids and the type of valve is specified by the total number of electrodes in it. Thus a single grid valve is called a *triode* because it has three electrodes: cathode, anode and one grid. A *pentode* has three grids.

Semiconductors (e.g. transistors): These have almost entirely replaced thermionic devices. Their action is more difficult to explain, but briefly it can be said that there is no heated electrode, they are much more efficient and vastly smaller than their thermionic counterparts, and have a longer life.

There is a belief in some quarters that for audio purposes valves have better sound-handling characteristics. There is little evidence to support this view.

Symbol for a rectifier or diode
Conventional current flow is in the direction of the arrowhead

Rectifiers and diodes: So what's the difference? Both allow current to flow in only one direction. The term *rectifier* is a general one and could, in principle, apply to any device that has the property of allowing current flow in one direction only. A clever switching unit that operated only on one half of an a.c. cycle could be called a rectifier. Also, there is often an implication that a rectifier works at the supply mains frequency.

The word *diode* is usually taken to mean a (possibly quite small) semiconductor component, although *thermionic* diodes are included in the category.

Semiconductor diodes have a small leakage current in the non-conducting direction. Also, as has been mentioned before under a.c. meters, the relationship between 'forward current' and voltage is not linear, especially at low voltages and currents.

Transistors: These have been mentioned above. The current flowing from one semiconductor material to another is modified by the current flowing into, or the voltage applied to, a separate layer of material between them. To some extent, then, they are a little like valves. However the current emitting part is called the emitter (not cathode), and this current flows to the collector (equivalent to the anode in a valve) while the controlling layer is known as the base.

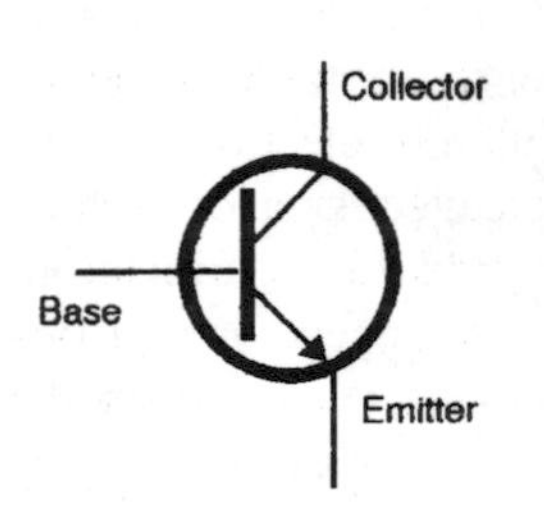

Symbol for a transistor

Field effect transistor (FET): A form of transistor in which the current flows through a narrow channel. The equivalent of the emitter in an ordinary transistor is called the source and the current flows to the drain. A voltage applied to the gate controls the current in the channel. A FET is rather like a thermionic valve in its working. One characteristic is that it has a very high input impedance: in other words it is effectively only a voltage which is applied to the gate. The current flowing into the gate is negligible, unlike ordinary transistors where it is really the current flow into the base which controls the current flowing from emitter to collector.

MOSFET (metal-oxide-silicon field effect transistor): A type of FET in which silicon is used to insulate the gate.

LED (light emitting diode): A semiconductor diode which has the property of emitting light. These are very much more efficient than filament lamps in that for a given light output they consume much less power. At the present time, however, they cannot compete with filament lamps in the production of light although sets of around six or eight are powerful enough to be used for bicycle rear lights. They are much used as indicator lights where their smallness and long life, as well as low power consumption, make them much more suitable than filament bulbs for many purposes.

3.2 Miscellaneous

Integrated circuit (IC): A complete circuit in one package, as opposed to the wiring up of *discrete* (separate) components. The term *chip* is commonly used for ICs. With modern technology extremely complex circuits can be produced in one small IC. LSI (large scale integrated (circuit)) and VSLI (very large scale integrated (circuit)) indicate degrees of complexity and size of the device.

Printed circuit board (PCB): A flat board of insulating material, typically from a few centimetres up to 10 or 20 cm across, produced with the wiring of a circuit printed on one or both sides with a conductive film. Small holes allow components to be inserted in their correct places and soldered in.

PCBs are mostly used in manufacturing of equipment where the numbers to be produced are not likely to be sufficient to justify the cost of producing ICs. (The latter can be very cheap when production runs are going to be of the order of thousands!)

Operating voltages: Most semiconductors (including ICs) operate at quite low voltages (up to perhaps five), as distinct from thermionic devices such as valves which generally need high voltages – typically from tens of volts up to maybe thousands.

4

Radio transmission

4.1 Radio waves

Radio waves are described as being *electromagnetic* in character. This means that they have an electric (electrostatic) component and a magnetic component. This arises from the fact that they are generated whenever electrons are accelerated or decelerated. And since electrons have an electric charge, and a flow of electrons (a current) results in a magnetic field, it's not hard to accept the fact that the waves produced are electromagnetic in character. For example, the waves used for radio and television transmission are electromagnetic waves and they are created by producing a high frequency current in an aerial. Now this means that the electrons which constitute the current are changing speed and direction constantly – being accelerated and decelerated constantly – and radio waves are emitted from the aerial.

In fact electromagnetic waves are far more all-embracing (almost literally!) than just the example of radio and television waves. Light, including ultraviolet and infra-red rays and X-rays are all electromagnetic (e.m.) waves. They differ only in their wavelength. The next section goes into this a little further.

4.2 The electromagnetic spectrum

The table below gives the wavelengths and the corresponding frequencies of the complete (as far as we know!) range of e. m. waves. Note that the frequency and wavelength are related by the formula

$$c = f\lambda$$

In this c is the velocity of the wave, which in the case of e.m. waves is very close indeed to 300 000 km/second, f is the frequency in Hz and the Greek letter l ('lambda') represents the wavelength in metres. (This same formula, but with a different velocity applies to sound waves. We shall meet it again!)

Part of spectrum	wavelength	Frequency range
Radio waves	centimetres to 10^5 m	Up to 3 x 10^8 Hz
Infrared	millimetres to 10^{-6} m	3 x 10^8 Hz to 3 x 10^{11} Hz
Visible	around 5 x 10^7 m	around 10^{12} Hz
Ultraviolet	around 10^{-8} m	around 10^{13} Hz
X-rays and gamma rays	less than 10^{-9} m	above 10^{14} Hz

Figure 4.1 shows the elements of the table above. Two important facts about the spectrum are:

1. The visible spectrum is actually a very narrow band of frequencies: the ratio of the highest (violet light) to the lowest (deep red) is less than 2:1, with yellow coming near the middle. When dealing with light it's usual to use wavelengths rather than frequencies and the most convenient unit is the nanometre, nm.

   1 nm = 10^{-9} m or 1/1000 mm.

2. Radio waves are divided up, for convenience, into sections. These are:

   VLF (very low frequency): The wavelengths here are extremely long and range from 10 000 m to 100 000 m.
   LF (low frequency): From 1000 m to 10 000 m. This covers the broadcast Long Wave band.
   MF (medium frequency): Changing our units, this band is centred on about 1 MHz and covers the 'medium wave' band on a domestic radio.
   HF (high frequency): Around 10 MHz and into the 'short wave' region.

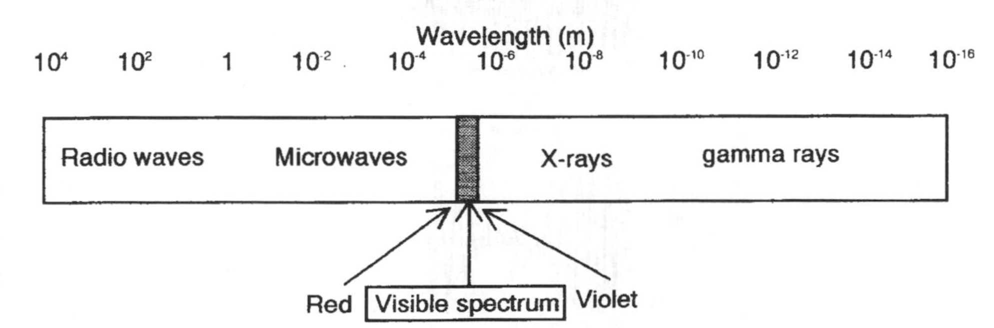

Figure 4.1 The electromagnetic spectrum

VHF (very high frequency): Around 100 MHz. In the UK vhf radio is from about 88 MHz to 110 MHz.

UHF (ultra high frequency): Around 1000 MHz (1 gigahertz). Television broadcast frequencies come into this range.

SHF and **EHF (super- and extra-high frequencies)**: These go up to 100 GHz and beyond.

(Microwave cookers operate in the SHF/EHF frequencies, using radio waves of about 10 cm.)

4.3 Amplitude modulation (a.m.)

Radio waves, and indeed all the other kinds of e.m. waves, are really only of interest to us if they are useful. With radio waves this means making them able to carry information, whether it be speech, music, pictures or any other form of communication. And this involves causing some characteristic to vary in a controlled way. We call this *modulation*.

There are several methods of modulating a radio wave and we will deal with the two most important, giving only a brief mention of others.

In our first system, a.m., the amplitude of the wave is varied to follow the shape of the modulating signal, such as the output of a radio studio. Figure 4.2 illustrates this.

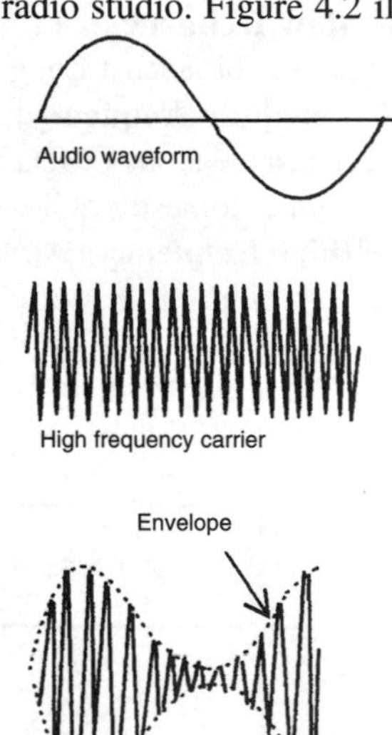

Figure 4.2 Amplitude modulation

The original wave is called the carrier and its frequency is therefore known as the *carrier frequency*, usually denoted by f_c, and the frequency of the modulating signal is represented by f_m.

An important aspect of a.m. (and all other modulation systems, as it happens), and one which is perhaps not obvious at first sight, is that the modulation process creates other frequencies, called *sidebands*. These are frequencies which are symmetrical about the carrier frequency and depend on both f_c and f_m. For example, take a carrier frequency of 198 kHz (which happens to be the frequency of BBC Radio 4 in the long wave band). If an audio signal of 1 kHz is used to modulate the carrier then there is a mixture of the following:

the carrier frequency f_c
the sum of f_c and f_m 198 +1 = 199 kHz,
and the difference, $f_c - f_m$ 198 – 1 = 197 kHz.

The sum frequency, 199 kHz in this case, is called the *upper sideband*, while the difference frequency, 197 kHz, is called, not surprisingly, the *lower sideband*. The sidebands, in this simple example, cover a spread of 2 kHz and this is known as the *bandwidth*.

One disadvantage of using the long wave (LF) bands is that relatively few channels can be accommodated because of the bandwidth. For just acceptable broadcast speech and music we need to transmit audio frequencies of at least 6 or 7 kHz (and that's a very long way from hi-fi but listeners to medium and long wave radio have accepted these limits fairly cheerfully for a long time!). Consequently the sidebands must be in the region of 12 to 14 kHz. Taking our 198 kHz carrier as an example, when used for broadcasting it then 'spreads' from about:

198 – 7 = 191 kHz to 198 + 7
= 205 kHz.

This is quite a substantial slice of the available frequency range of a domestic radio, which typically goes from about 120 to 300 kHz.

In the medium wave (MF) band things are a little better but not much. Taking a frequency of 1 MHz, this will spread because of sidebands from 993 to 1007 kHz.

Notice that we have allowed 7 kHz for the upper audio frequency allowed in a.m. broadcasting. Actually international agreements allow up to 9 kHz, but many broadcasters limit their upper audio frequency to much less, simply because if the sidebands of different stations overlap there are likely to be disturbing whistles and other interfering noises, much of which is readily observable when listening to a.m. broadcasts, particularly from the weaker stations after dark, when these radio frequencies can travel long distances.

Broadcasting on a.m. has several drawbacks: one we have just mentioned, the likelihood of interference from distant transmitters. Another is that natural, so-called '*static*' interference from thunderstorms, together with man-made electrical disturbances caused by switching equipment on or off, is an amplitude-type of effect. This means that a.m. broadcasts can be quite badly affected by this sort of interference. A thunderstorm can be many hundreds of miles distant and still contribute to a kind of 'mushy' background noise, and, nearer at hand, central heating switching in and out, and the opening of a fridge door even, can add to the trouble.

4.4 Frequency modulation (f.m.)

In f.m. it is not the amplitude of the carrier which is caused to vary, it is the frequency. Figure 4.3 shows this.

Notice that the amplitude is constant. The big advantage of f.m. over a.m. is that it is a system which is much less prone to interference, whether man-made or natural, for the reason that the effects tend to be amplitude effects. These might occur and affect the f.m. wave but receivers are designed to disregard such things, being concerned only with the frequency.

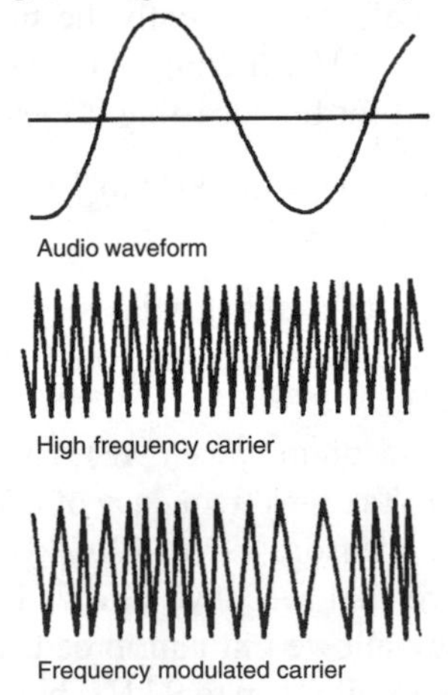

Figure 4.3 Frequency modulation

As with a.m. there are sidebands, and these can be extensive. In theory they extend almost infinitely on either side of the carrier. To begin with there is the matter of the *frequency deviation* – the amount by which the carrier frequency is allowed to change. This can vary with different systems but for broadcast radio the maximum deviation is typically ±75 kHz. This represents the maximum amplitude of the audio, or other, modulating signal. Then the modulating *frequency* has to be allowed for, and a reasonable practical figure for the sidebands is to add the highest audio frequency to the deviation frequency. Now, we allowed a vague figure of about 7 kHz for a.m. but, because f.m. is capable of much better quality in every way, we can allow 15 kHz for the highest audio modulating frequency.

This makes the *total frequency* deviation the sum of the highest audio frequency *plus* the allowable carrier deviation:

$$(75 + 15) = \pm 80 \text{ kHz}$$

or a total bandwidth of about 160 kHz.

There is no way in which that would be acceptable in the LF or MF bands! One f.m. broadcast would just about fill the LF band, and there wouldn't be space for many in the MF band. So the answer is to put f.m. broadcasting into the VHF region.

Taking 100 MHz as being about the middle of the VHF band then the bandwidth would be 80 kHz on either side of 1 MHz, that is:

1000 – 80 kHz for the lower sideband, and
1000 + 80 kHz for the upper sideband.

In other words, 920 to 1080 kHz, which is much more reasonable. The VHF broadcast band goes from 88 to 110 MHz, a range of 22 MHz. In principal 22 MHz ought to be able to accommodate about 140 broadcast channels.

Now let's see the advantages and disadvantages of f.m. compared with a.m.

1. As we have already said f.m. is much less affected by electrical interference of an a.m. kind – whether man-made or natural. Thunderstorms, for instance, have almost no effect on reception.
2. It's possible to transmit much higher audio frequencies than a.m. and this comes about, at least partly, because of using the VHF bands. In doing this there are simultaneous pros and cons.

The maximum range of VHF transmitters is generally not much more than line-of-sight – perhaps 30 or so miles (50 km), maybe a little more in flat terrain. In certain atmospheric conditions this can be much greater. This is because VHF radio waves don't bend round the Earth as MF and LF waves can do. So, no matter how powerful the transmitter, something in the region of 30 miles is about the greatest distance for coverage. The big plus for this is that distant, i.e. more than 50 to 100 miles, transmitters, even if on the same frequency, are unlikely to cause interference.

The drawback is that many transmitters are likely to be needed to give decent coverage of a country, and this is made worse by the presence of hills which act as effective screens! In the UK there are main, high-power, transmitters which can cover large areas of country where there are no significant hills, which is the case in much of the south of England and the, Midlands although even there low-power transmitters can be found for covering local areas of poor reception, particularly towards the limits of the main transmitter's range.

In hilly regions, such as Scotland, Wales and much of northern England it becomes necessary to have a great many of these low-power transmitters, for example at heads of valleys and anywhere that high ground prevents reception from the main transmitters.

4.5 Other modulation systems

Modulation systems can be broadly divided into two general categories:

1. **Analogue**, in which the original carrier is modified in a way that is proportional to the original modulating signal. a.m. and f.m. can both be regarded as analogue modulation process because in the case of a.m. the amplitude is following the shape of the audio signal and with f.m. a graph of the frequency would be the shape of the audio signal;

   Among other analogue methods we can list the two most important:

   (a) *Suppressed carrier*. In an a.m. system the sidebands carry all the wanted information, and yet the carrier is still there. There can be significant savings in the power produced by a transmitter if the carrier is removed from the transmitted signal. There is a problem, though, at the receiver. A small signal of the same frequency as the transmitter needs to be generated in order to covert the a.m. into a usable output. In stereo radio transmission a trace of the carrier is used to make it easy for the decoder to work.

(b) *Single sideband, suppressed carrier.* Since the upper and lower sidebands are exactly equivalent to each other a further saving in transmitter power is possible if only one sideband, and no carrier either, are transmitted.

2. **Digital modulation**, in which the signal to be handled isn't put on to a carrier. Instead complex processes convert it into what is essentially a numerical version of the original. We shall deal with this in more detail later. At the moment we shall do no more than state that the final signal (which may be recorded or possibly used to modulate a carrier) consists of *pulses* which are all the same height and width but which carry numbers representing very small sections of the original signal. This is more commonly known as *digital audio* or *digital video*, or less often, as *pulse code modulation* (PCM).

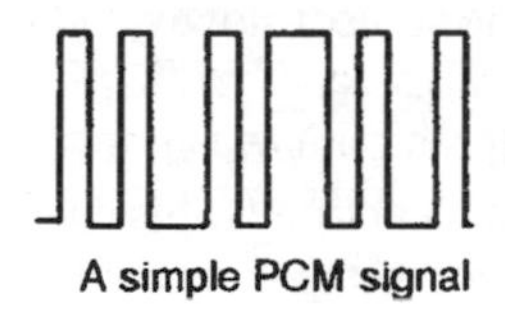

A simple PCM signal

A variant of PCM is *pulse width modulation* (PWM). In this the pulses are all the same height but they vary, obviously, in their width.

5

Decibels

5.1 What are they?

Let's say at the outset that decibels can be very confusing. This is at least partly because they are very flexible units that can be used in a variety of applications and sometimes one isn't very sure what particular application to use in what set of circumstances. However we shall take the subject slowly and in easy stages.

Decibels are units of comparison

To begin with, a decibel (dB) is a unit of *comparison*: it isn't a unit in the sense that metres or kilograms are units. Instead it can be used to compare two powers, two voltages, two sound pressures, and so on.

Its original application was to give a measure of the relative values of two powers, but the decibel is not a simple ratio – it is the *logarithm* of a ratio – and that's perhaps why people coming across decibels for the first time are immediately dismayed! However, we don't need to go into the mathematics of logarithms, and in any case very modestly-priced calculators can quickly handle the numbers. Let's begin by seeing why logarithms can be useful.

Suppose we have an amplifier with a small power going in, say 1 milliwatt (1 mW) and a power coming out of 30 W. (Later we'll see that it's unusual in audio work to be concerned about the power going into an amplifier. More usually it's the voltage. However, let's keep things as uncomplicated as we can for the time being.)

The ratio of the power out to the power in is:

$$30/0.001 = 30\,000.$$

This is a large number and in some applications we could be dealing with much larger ones still. The logarithm of 30 000 is 4.477 – a small number. If this is puzzling read the panel on the next page. If not, skip the panel!

Now suppose we have two amplifiers, one following the other.

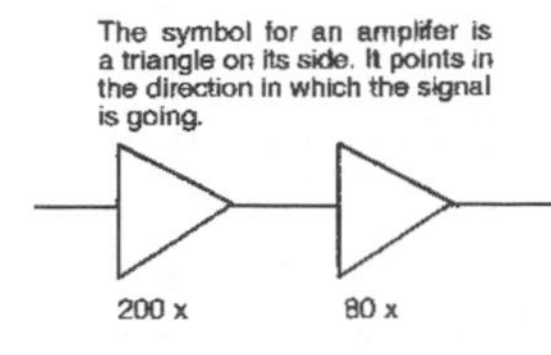

The symbol for an amplifer is a triangle on its side. It points in the direction in which the signal is going.

The first one amplifies the power by 200 times, the second by 80 times. The total amplification is 200 x 80 =16 000, which isn't a very difficult calculation, admittedly. But, if we are told that the *gains* of the two amplifiers are 23 and 19 dB respectively then the total amplification is 23 + 19 = 42 dB which could, in many circumstances, be a much easier bit of arithmetic.

Now, to find the logarithm of 30 000, which we used above, because this is 3 x 10 000, we add the logarithm of 3 (0.477) to the logarithm of 10 000 (4) making 4.477.

Never mind about how logarithms are defined. Let it be sufficient to say that every number has a logarithm ('log' for short) and this log can be found from all but the cheapest pocket calculators. The important point is that adding logarithms is equivalent to multiplying the original numbers, and subtracting is equivalent to dividing them. The reader may wonder what use this is if one can do the same functions on a calculator without using logarithms. A fair point! However, as we shall shortly see, this is a great help because it means we can add or subtract decibels and this can simplify any calculations enormously.

A few simple logs, and probably almost all that we are going to need in this book, are:

Number	*Logarithm*
1	0
2	0.301
3	0.477
4	0.602
5	0.700
6	0.778
7	0.845
8	0.903
9	0.954
10	1
100	2
1000	3
10 000	4

5.2 A definition

In mathematical terms the number of decibels is defined as:

10 x log(power ratio)

Most of the time in audio work we are concerned with voltages rather than powers.

Now we said in Chapter 1 that power in watts was proportional to $(\text{voltage})^2$, so that:

$$P = V^2/R$$

To square a number we multiply the logarithm by 2 and that means that if we are using voltages the number of decibels change becomes:

20 log(voltage ratio).

All this might seem rather complicated, but maybe an example or two can help make things clearer.

Example. the output voltage of a typical microphone is in the region of 0.5 mV. How many decibels must an amplifier provide to raise this to 2 V?

The voltage ratio is 2/0.005 = 4000.

Therefore the amplifier must have an amplification, or gain, of:

20 log 4000 = 20 x 3.602

= 72.04 dB.

(since log 1000 is 3 and log 4 is 0.602)

In fact it is almost certainly going to be accurate enough in practice if we gave the answer as 72 dB.

Quite often we need to work 'backwards' – we know the gain in decibels of a device (it might not be an amplifier: it could be a voltage-reducing circuit – an *attenuator*) – and we need to find the voltage either at the input or at the output.

It's usual to indicate gains with *positive* numbers of decibels, and reductions (or losses) with negative decibels. Thus, if we find a figure of +30 dB we know that an increase in level of the voltage or the power has occurred, similarly -10 dB tells us that there has been a reduction in level, for whatever reason. Usually,

Example. An amplifier has a gain of 60 dB. What is the input voltage if the output voltage is 5 V?

We can write:

$$60 = 20 \log(5/Vin)$$

This might seem rather complicated. In fact it isn't really, because we can divide both sides by 20:

$$60/20 = \log 3$$
$$= \log (5/V_{in})$$

The antilog of 3 is 1000, so

$$1000 = (5/V_{in})$$

or

$$V_{in} = 5/1000$$
$$= 5 \text{ mA (milliamperes)}$$

though, we don't put the + sign in except where there might otherwise be an ambiguity.

At this stage a table of decibels for different voltage ratios might be helpful.

The table is worth a certain amount of careful study and a few very useful points can emerge, not the least of which is that it becomes possible, with a little bit of practice, to be able to 'guesstimate' decibel values quite easily!

To begin with, doubling or halving a voltage ratio represents

Voltage ratio	Decibels	Voltage ratio	Decibels
1/1000	-60	3	9.5
1/500	-54	5	14
1/250	-48	10	20
1/100	-40	20	26
1/50	-34	30	29.5
1/10	-20	100	40
1/5	-14	200	46
1/3	-9.5	500	54
1/2	-6	1000	60
1/1	0.0	10 000	70
2	6.0	100 000	80

a change in the decibels of 6: +6 if it's a doubling, –6 if it is a halving.

Secondly, if the ratio is changed by a factor of 10 then 10 decibels added (or subtracted). The last lines of the table show that if a voltage ratio is 100/1, that represents 20 dB; if it's 1000/1 – a factor of 10 increase – then the 20 dB becomes 30 dB, and so on.

Example. How many decibels correspond to a voltage increase of 50 times?

50 doesn't appear in the table, but 100 does and that's 40 dB. 50 is half of 100 so all we do is to subtract 6 (because halving or doubling a voltage is equivalent to a change of 6 dB). Taking 6 from 40 makes 34 dB.

All this might seem difficult stuff! However we shall meet decibels again, and again, and again! And the reader will find that before long he or she will become quite used to dealing with them. There's no need to worry about the mathematics. We've outlined that aspect here for the sake of completeness but much of the time calculations don't come into it – except to add or subtract numbers.

A final point in this chapter. Readers who have already been involved with sound in a professional way will possibly have met decibels with additional letters or subscripts, such as dBu or dBA. We'll deal with these as they arise.

6

Sound waves

6.1 Types of wave

There are broadly two kinds of wave in nature. First there is the familiar type of wave such as we can see in water where the *medium* visibly rises and falls. Such waves are known as *transverse* waves because the motion of the individual particles (water molecules, for instance) is an oscillation at right angles to the direction in which the wave is travelling. Interestingly, electromagnetic waves (radio waves, light, X-rays and so on) are classed as transverse, although it isn't easy to say what the particles are since such waves can travel though a vacuum perfectly easily.

Sound waves, on the other hand, are waves of compression and rarefaction. The individual particles – in reality air molecules – oscillate in the same direction as that in which the wave is travelling. Figures 6.1 and 6.2 illustrate the difference between the two kinds of wave.

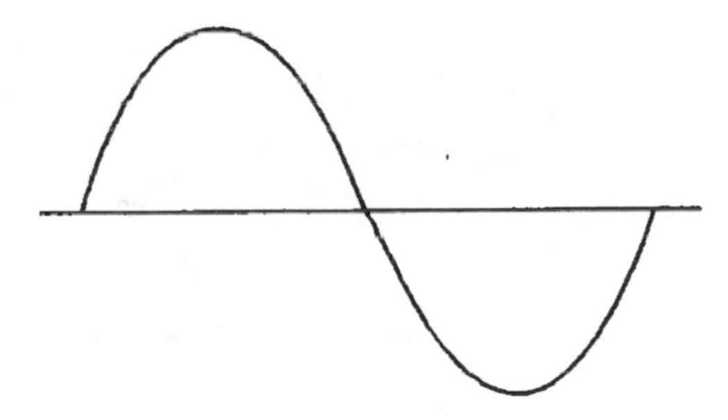

Figure 6.1 A transverse wave

Figure 6.2 A longitudinal wave

Figure 6.3 represents a sound wave. At a compression the air molecules are more closely packed and the pressure is relatively high; at a rarefaction there is a region of low pressure. Waves in which the particles oscillate in the same direction as the direction of the wave are called *longitudinal* waves. With sound waves the compressions cause the eardrum to be moved in, very slightly, while the rarefactions cause it to be drawn out, also very slightly. The eardrum movements are so small that with a just detectable sound the movements of the eardrum are comparable with the diameter of a hydrogen atom, and that is the smallest atom there is!

Diagrams like Figure 6.2 are difficult to draw, and also they don't convey at all well the 'shape' of the sound waveform. Consequently we almost invariably use a graph of some aspect of the sound wave, usually pressure, and the pressure graph of a pure sound wave would look like Figure 6.1. The horizontal straight line then represents atmospheric pressure, the region above the straight line is representing pressures higher than atmospheric, and so on.

We must emphasize, though, that the variations above and below atmospheric pressure are very small, even in the loudest sounds. The pressure variations in even a very loud sound wave are very much less than even one millionth of normal atmospheric pressure!

In future then we shall use graphs of sound wave pressure as in Figure 6.3.

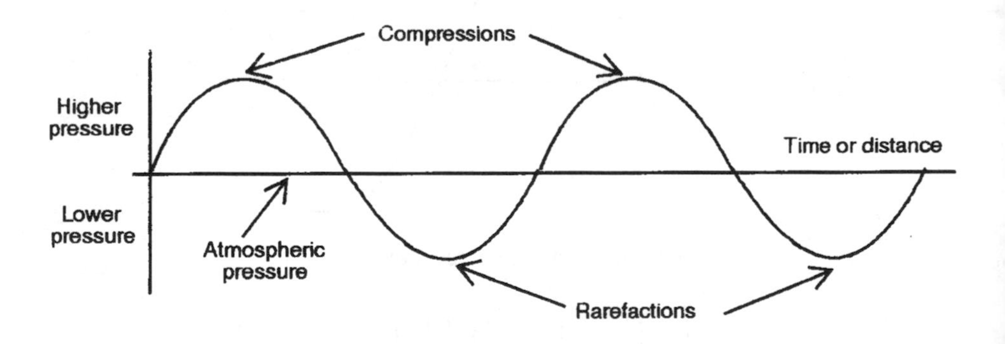

Figure 6.3 Representation of a sound wave

6.2 Sound wavelengths and frequencies

Let's take frequencies first. The normal adult human ear can detect frequencies in the range from about 16 Hz at the very lowest to about 16 kHz at the highest. There is, though, a fair amount of variation in these numbers. To being with, at the lower end it's very difficult to decide when a sound ceases to be a perceptible note and when it starts to be some sort of flutter in the ear. Some writers give 20 Hz as the lower limit – we shall take 16 Hz, not because it's necessarily more accurate but simply for the reason that it's easy to remember if we take 16 kHz as the upper figure. But even that is a rather suspect value! Up to the age of late teens or very early twenties people can typically hear up to about 20 kHz – possibly more. However from then on the upper limit drops off, according to one estimate by about 1 Hz a day! There do seem, though, to be wide differences. Some people at the age of fifty can hear little above 10 kHz, others can get to their sixties and still hear perfectly well up to 12–14 kHz. Despite all that, it's reasonable to take 16 Hz to 16 kHz as an approximate hearing range for the average person.

Now, to turn to the wavelengths involved, and that's going to be important later because, as we shall see, the behaviour of sound waves in a room or studio can depend critically upon the relationship between the sizes of obstacles and the sound wavelengths.

Before we can work out wavelengths for different frequencies we need to know about the velocity of sound waves, and then we can use the simple formula we met earlier relating velocity, frequency and wavelength:

$$c = f\lambda$$

The velocity of sound varies with temperature, increasing as the temperature rises. At normal room temperatures c is 340 m/s and we shall stick to this figure throughout the book, as the variations aren't normally very important, except for players of wind instruments who find that things like trumpets aren't at their correct pitch when they are cold. (To be quite accurate we will add that at 0°C c is 330 m/s and increases by about 0.6 m/s per °C)

If we put c = 340 m/s in the formula we can (to save the reader having to do the calculations) produce the table given on the following page.

Frequency (Hz)	Wavelength (m)
16	21.25
30	11.33
50	6.8
100	3.4
500	0.68
1000	0.34*
2000	0.17
5000	0.068 (6.8 cm)
10 000	0.034 (3.4 cm)
16 000	0.021 (2.1 cm)

* It is useful to remember this line of the table; a sound wave of frequency of 1 kHz has a wavelength of one third of a metre to within a quite reasonable degree of accuracy.

Furthermore it's often quite good enough in practice to use that fact as a basis for rough calculations of other wavelengths. For example, starting with 1 kHz (l = 1/3 m, approximately) we can then say:

500 Hz has a wavelength of about 2/3 m
340 Hz .... 1 m (accurately)
100 Hz .... 3 m
50 Hz .... 6 m

and so on, with a similar sort of reasoning for frequencies above 1 kHz: 2 kHz has a wavelength of about 16 cm; 5 kHz, 7 cm.

It may be comforting to know that this kind of accuracy is often quite good enough in the practical world.

6.3 Sound waves and obstacles

Everyone is familiar with echoes: the reflection of sound from buildings, and the interested observer can notice, for example, an 'echoey' effect in woods. What is perhaps not obvious is that for reflection to occur the reflecting object must be bigger than the wavelength of the sound waves. We had better be a little more exact – the *dimensions of the object at right angles to the direction of the wave must be greater than wavelength for there to be significant reflection.* Figures 6.4 and 6.5 illustrate this.

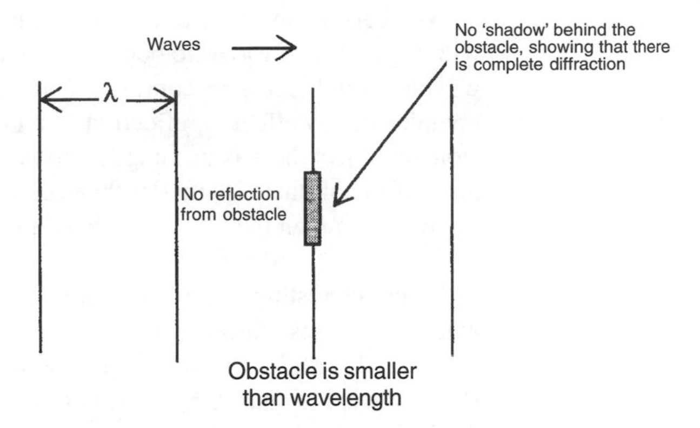

Figure 6.4 Reflection – λ is larger than the obstacle

What we have just said applies to all waves, not just sound waves. It's very instructive, for instance, to stand on a jetty and watch how waves in the sea behave. Long waves can be seen to be reflected only from, say, the hulls of boats. Waves a metre or two in length go past small mooring buoys as though they did not exist. Light waves are affected by the same rules, but in their case the wavelengths are so small (around 500 nm or about 0.5 of a thousandth of a millimetre) that all everyday objects are vastly larger than λ so that reflection occurs in every perceptible case. Indeed, we can only see objects because of the reflection of light waves from them so it follows that we can only perceive objects larger than λ !

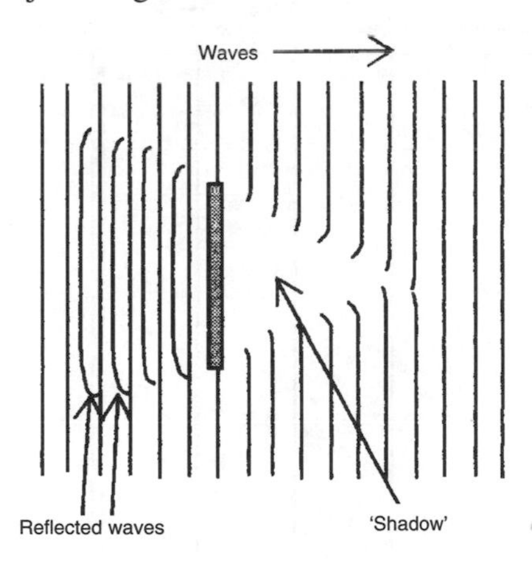

Figure 6.5 λ is small compared with the obstacle

Another, and in a way related, effect is *diffraction*. This deals with the ability of waves to bend round an object – but again the wavelength/object size comes into it. Diffraction is really a complementary effect to reflection. If all the waves are reflected from an object there is nothing left to bend round. On the other hand, if the obstacle is smaller than the wavelength and there is no reflection then the waves are bending, or 'diffracting' round it.

Some interesting and important aspects of diffraction occur when waves pass through an aperture. If the aperture is large (compared with λ) the waves emerging do so in a relatively straight beam (Figure 6.6 (left)) but with a small aperture then the beam spreads widely (Figure 6.6 (right)).

An immediate application of these effects is in loudspeakers. A loudspeaker, certainly if of the conventional type, is really an aperture in a box with a sound-producing device behind the aperture. On a reasonably-sized monitoring speaker the aperture – hole – is typically around 30 cm across. When the sound being generated by it has wavelengths much more than 30 cm – say about 3 m, the corresponding frequency being around 100 Hz – then the sound spreads over a wide angle. If, though, the frequency is high, say about 3 kHz, the wavelength being in the region of 3 cm, then the sound comes out in a fairly narrow beam. A consequence of this is that a listener is unlikely to receive the high frequencies properly unless he or she is sitting more-or-less on the axis of the loudspeaker.

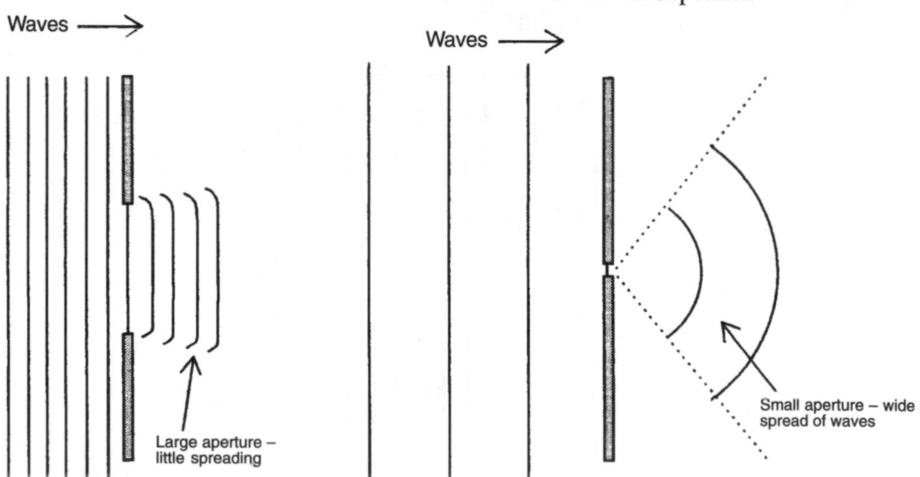

Figure 6.6 Diffraction effects at an aperture

Designers try to broaden the high-frequency response as much as they can, and we shall see in a later chapter how this is achieved, at least partly.

6.4 The inverse square law

Everyone knows, as a matter of common sense, that sound waves, or any other kind of wave for that matter, become 'weaker' the further they travel from the source. Here is the place to be a little more specific about the effect. But first we must introduce the idea of *intensity*. There is a temptation, perhaps, to regard this as being the same thing as *loudness*. Loudness, though, in scientific terms, is somewhat different – we shall go into this later.

Here we must adopt the true definition of intensity as being the amount of wave power passing through, or falling upon unit area. We can use the watt as the unit of power so the units of intensity are:

watts/square metre or w/m^2.

Perhaps we should point out here that in dealing with sound an intensity of 1 w/m^2 is far greater than anything we normally meet. Small fractions of a microwatt per square metre are more likely!

To begin with we must assume that the waves spreading out from a source are free to do so in every direction, so that there are no obstructions of any kind. This is, of course, a somewhat unrealistic situation – we'd need to imagine a loudspeaker suspended some distance below a balloon to approach these conditions. Unreal though the normal state of affairs might seem, there are useful approximations that we can make to be realistic.

It doesn't take much imagination to see that if we have an isolated source of waves which are free to travel out in all directions, then if we could see a single wave it would look like a sphere. A moment later this sphere would have expanded, and Figure 6.7 illustrates this.

If, having gone double the distance, the area through which the waves are passing has doubled then the intensity, which we represent by *I*, must have quartered. This leads us to a formula for the inverse square law:

$$I \propto 1/d^2$$

where *d* represents the distance and μ means 'is proportional to'.

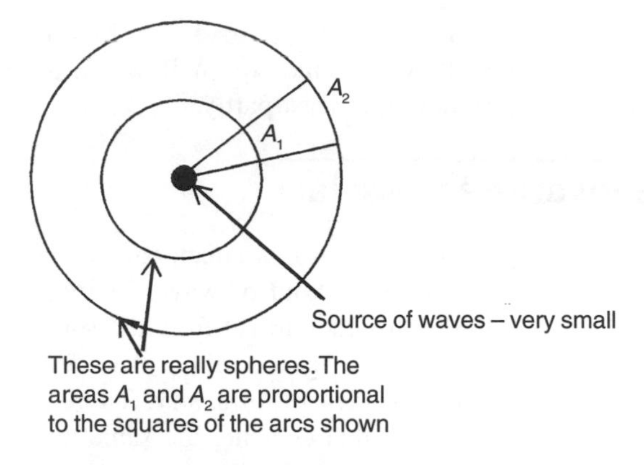

Figure 6.7 The inverse square law

Every doubling of the distance reduces the sound level by about 6 dB (usually a little less in practice)

So, if we have a certain intensity of sound wave at, for example, 5 m then at 10 m the intensity will be a quarter, at 20 m (4 times the distance) it will be one sixteenth, and so on.

Pausing for a moment to translate this into decibels: for each doubling of the distance the intensity is quartered, so the number of decibels reduction will be 10 log 4 which is 10 times 0.602 (or 0.6 is near enough), which comes to 6 dB.

This, as we've said, isn't strictly accurate except in special circumstances, but it can be a very useful approximation.

6.5 Sound in pipes

There are very many musical instruments which are basically lengths of pipe with a noise generating device at one end. The pipe may be wood or metal; it may be straight or coiled; it may be of constant diameter or it may be conical. The noise generator can be the player's lips, or a vibrating reed or pair of reeds, or simply turbulence set up when an air stream strikes a suitably shaped surface. But whether we are thinking of trumpets, flutes, organ pipes or whatever they are all essentially some form of pipe. And, surprisingly perhaps, a room or studio can be thought of in the same way, at least to begin with. Figure 6.8 shows a pipe closed at one end.

If sound waves enter from the open end they will be reflected back from the closed end and if it so happens that the frequency is such that it is related to the time it takes for the sound waves to travel down the pipe and back then interesting things happen. The first is that they can be *reflected back again from the open*

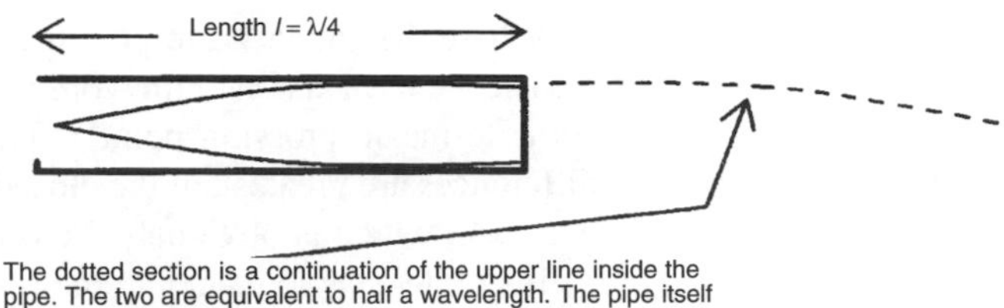

Figure 6.8 Pipe closed at one end

end! This may seem surprising. The fact is that sound waves travel inside a pipe slightly slower than they would in the open air – because of friction effects at the walls of the pipe. At the junction between the pipe and the open air there is what is termed a *mismatch* in the conditions, and some (possibly only a small proportion) of the wave energy is reflected.

We have, then, a state of affairs in which some of the sound wave is bouncing to and fro inside the pipe. If this rate of 'bounce' – or frequency of oscillation – happens to coincide with the frequency of the sound wave being generated outside we have a condition known as *resonance*.

Resonance is actually a very common phenomenon – we find it in all kinds of situations and not just with sound waves in pipes. For example it's quite common to get a rattle in a car at certain speeds (unless the car is very new and possibly expensive!). There is an interesting effect in the lower parts of the River Severn in the UK where the rise and fall of the tides is several times higher than elsewhere because that section of the river happens to have a natural frequency of oscillation which is the same as the frequency of the tides.

An important feature of any resonating system is that the amplitude of the vibrations builds up to be greater, possibly many times greater, than the amplitude of the *exciting* vibration. When the amplitude is its maximum – having built up from nothing – the energy being lost by, for example, friction, is equal to the energy being supplied by the exciting vibration.

Figure 6.8 shows the simplest form of resonance in a pipe. The solid lines are actually graphs of the sound wave pressure. Notice that the pressure is greatest at the closed end of the pipe – as you might expect. It is zero (that is, no different from atmospheric pressure) at the open end, which again makes sense because we couldn't have two different pressures there!

At this point we must introduce some terminology.

Where the sound wave pressure is zero, as at the open end, we use the term *node* (Latin word meaning 'knot' but which has come to mean 'crossing point'). The place where the pressure differences are greatest, at the closed end of the pipe, we call an *antinode*, which is obviously the opposite to a node.

It's quite easy to calculate the resonant frequency, given the length of the pipe:

Suppose the length of the pipe, l, is 1 m. Then the wavelength is four times this – 4 m. We can work out the frequency from $c = f\lambda$:

$$
\begin{aligned}
340 \text{ m/s} &= f \times 4 \text{ m} \\
f &= 340/4 \\
&= 85 \text{ Hz}
\end{aligned}
$$

It's possible to have resonances at other, higher, frequencies. The important rules are:

1. There must be an antinode at the closed end.
2. There must be a node at the open end.

Figure 6.9 shows a higher mode of resonance. It follows the two rules we have just laid down and the length of the pipe is 3/4 of the wavelength.

In fact one can keep on with this, following the two rules, and after a short while a basic pattern emerges: a resonance can occur provided the length of the tube is an *odd number* of quarter wavelengths – 1, 3, 5, 7 and so on.

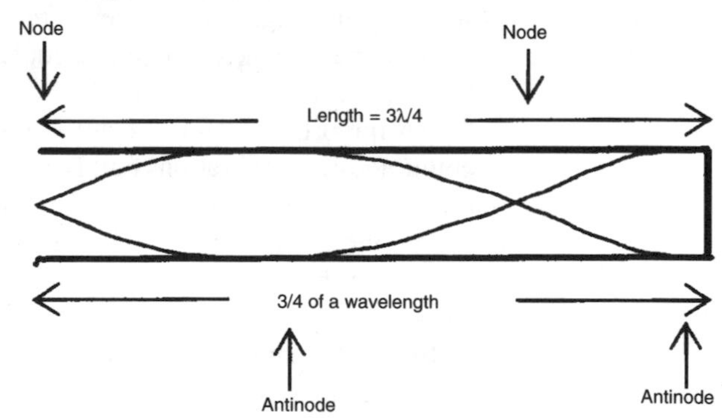

Figure 6.9 Pipe length is 3/4 of a wavelength

The case of a tube open at both ends is interesting. Here there must be an antinode at each end, as in Figure 6.10.

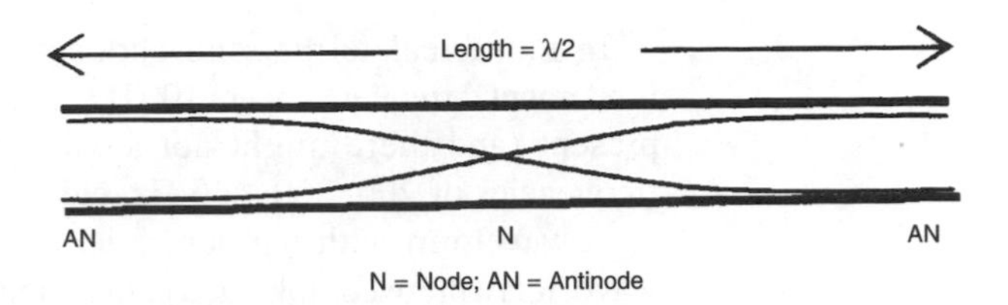

Figure 6.10 Pipe open at both ends

The reader should have little difficulty in seeing that the lowest resonant frequency corresponds to $l = \lambda/2$ and that other resonances can occur when $l = 2\,\lambda/2$ (or λ), $3\lambda/2$, $4\,\lambda/2$ ($= 2\,\lambda$), and so on. In other words a resonance can occur when the length of the tube is a whole number of half wavelengths.

What about a tube *closed at both ends*? This might seem silly because the reader may well ask how the sound gets in! Well, actually this is not as silly as it seems because the space between opposite walls in a room or studio can be thought of as a pipe closed at both ends! We shall return to this later.

6.6 Harmonics, etc.

We have seen that it is possible for a single pipe to be set into different modes of vibration, all of them resonant conditions. The same is true of vibrating strings, with the difference that we then have to think not of pressure nodes and antinodes, but of displacement nodes and antinodes. By 'displacement' we mean the amount of movement away from the rest position. This is sometimes a useful concept with sound waves – we shall find an application for it later – but in general pressure is the more convenient sound wave property.

However, the important thing at the moment is to realize that a vibrating object can basically resonate at a number of frequencies. From what we have said earlier in this chapter it should be reasonable to state that all such frequencies are *multiples of a basic frequency*.

The basic, or lowest frequency, is called the *fundamental* and all the others, which are simple multiples of the fundamental frequency, are called *harmonics*.

The lowest frequency present in a note is called the fundamental

We often denote the fundamental frequency by f_0 and then the frequencies of the harmonics can be written:

$$2f_0,\ 3\,f_0,\ 4f_0,\ 5f_0$$

and so on.

In numerical terms, and choosing casy numbers, if the fundamental frequency were 100 Hz then if there were harmonics present (and there might not always be!) they would have frequencies of 200, 300, 400 Hz, etc.

A waveform with harmonics is, unsurprisingly, not a sine wave. It is a more complicated pattern and is known as a complex wave. Figure 6.11 shows an example.

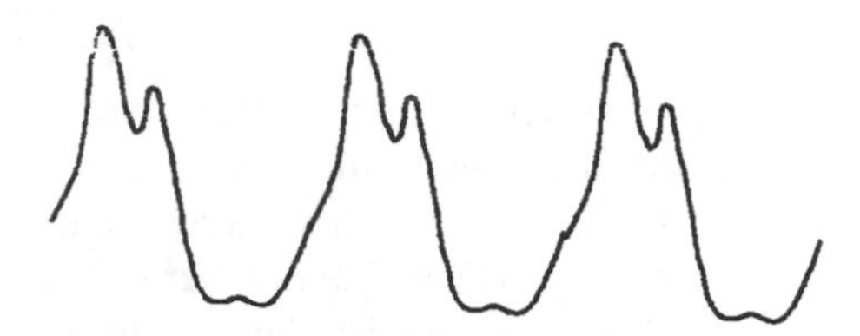

Figure 6.11 A complex wave

The proper definition of a complex wave is that it is a non-sinusoidal wave which is recurring. That means that each cycle, however complicated, is an exact replica of the preceding cycle. The last point is important. Figure 6.11 is a complicated wave but it is not a true complex wave. To the ear a complex wave is likely to be a reasonably pleasing sound. An non-recurring wave may be rather unpleasant.

6.7 What is a musical sound?

The statement that non-recurring waves may sound rather unpleasant brings us easily to the question of *sound quality*. In other words what is it that distinguishes the sound of, say, a flute from a saxophone? Sometimes the *pitch range* – highness or lowness of the note – gives some clues. Tubas are deep-sounding, piccolos are high-sounding. But there is much more to it than that.

In the last century experiments were carried out and it was found, even using quite primitive apparatus, that the notes of different instruments had very different *harmonic contents*. It was believed for a long time that this was the only factor which gave the ear information allowing it to identify the source of a sound, although even then it was realized that the harmonic structure changed for an instrument depending on the exact note it was playing.

Then, when tape recording became common in the 1950s, it was noticed that if a tape were played backwards the quality of the sound could change drastically – and yet the harmonic structures of the notes would not be affected.

Since then it has been found that virtually any sound, whether from a musical instrument or a human voice, has, for a brief fraction of a second at the start, additional frequencies which are not harmonically related to the fundamental frequency. They die away within a short time, possibly only a few milliseconds, and yet they appear to be vital in the recognition process. They are known as the *starting transients*. Indeed, if they are removed, or merely modified, the apparent tonal quality – we should use the term *timbre* (pronounced 'tarmbre') – of the sound can be greatly changed.

All this is of considerable importance in audio technology. Any microphone or loudspeaker must be able to reproduce the starting transients correctly, otherwise there cannot be satisfactory reproduction as judged by the ear – and the ear is, after all, the final judge, and probably the most critical and accurate one. When, in a later chapter, we come to deal with devices for automatic control of recording levels we shall see that the attack time of the devices, that is, the time that they take to start to function, is often chosen with special regard to the effect on the transients. A very rapid attack time, which might at first seem desirable, may affect adversely the starting transients of a sound and change their perceived quality.

7

The ear and hearing

7.1 The ear's response to sounds

We have talked about sound waves, and we have shown that they are longitudinal waves in the air, and that they travel in the air at about 340 m/s, and that the normal human can detect such waves if their frequencies lie between about 16 Hz and 16 kHz. But we have been making an unspoken assumption – that the ear *responds equally to all sounds in that frequency range*. This turns out to be quite untrue.

To begin with let us take what is called the *threshold of hearing* or *threshold of audibility* (the two mean the same thing). Experiments have been carried out with large numbers of people who have been asked to listen to a steadily quietening tone and then indicate when they can no longer hear the tone. The results are a graph like Figure 7.1.

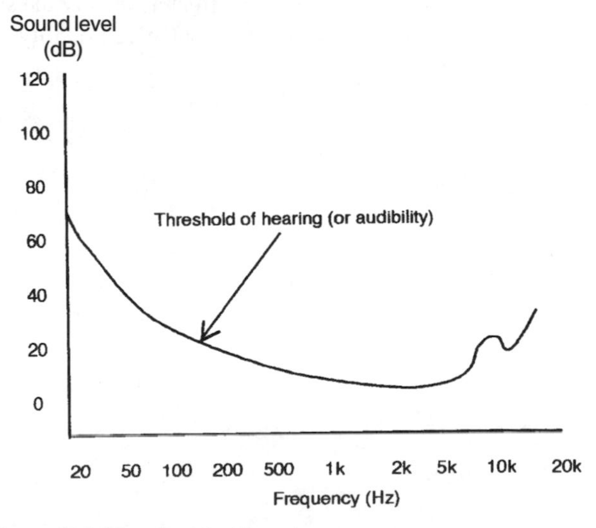

Figure 7.1 Threshold of hearing curve

The horizontal axis is frequency; the vertical axis is in decibels. We know that the decibel is a unit of comparison, so what in this case is being compared with what? Basically it doesn't matter very much. Any convenient sound pressure will do but to make all the numbers positive the *zero* is chosen to be the smallest detectable pressure – in other words near the lowest part of the curve.

Straight away one thing is very obvious: the curve is far from flat! We can summarize the shape in words like this:

1. At low frequencies the ear needs about 70 dB more sound level to make it work than it does at about 3 kHz.
2. The curve is at its lowest, meaning the ear is at its most sensitive, at around 3 kHz.
3. At high frequencies, 10 kHz and above, the ear is less sensitive but by no means as insensitive as it is at very low frequencies.

It is, of course, possible to convert the decibel indications on the vertical scale into sound pressures, in newtons per square metre or sound powers in watts per square metre. Taking pressures then at the lowest point on the graph the pressure is about 0.00002 N/m^2. (This is equivalent to a sound power operating the ear drum of roughly 10^{-16} W!) The figure of 0.00002 N/m^2 is taken as the zero (the *reference level*) for the decibel scale.

At the very low frequencies, where the sound level needed to cause a sensation in the ear is about 70 dB above the reference level, the pressure is about 0.06 N/m^2, some 3000 times higher.

We can go further than this in examining the ear's response. In the 1930s some researchers in the USA were the first people to produce *equal loudness curves*. Their work was repeated at the UK National Physical Laboratories using more modern equipment and techniques in the 1960s and the graphs produced there are known as the NPL curves. They are reproduced in Figure 7.2.

Basically the way these curves are produced is this: a person has a tone of particular frequency and loudness played to him or her. The frequency is then changed and the person is asked to make an adjustment so that the loudness is the same. The person carrying out the experiment can then note the new level of the sound in decibels and the process is repeated for another frequency. By doing this for many frequencies and at many levels the curves can be plotted. The ones shown are, of course, the result of such observations carried out with large numbers of

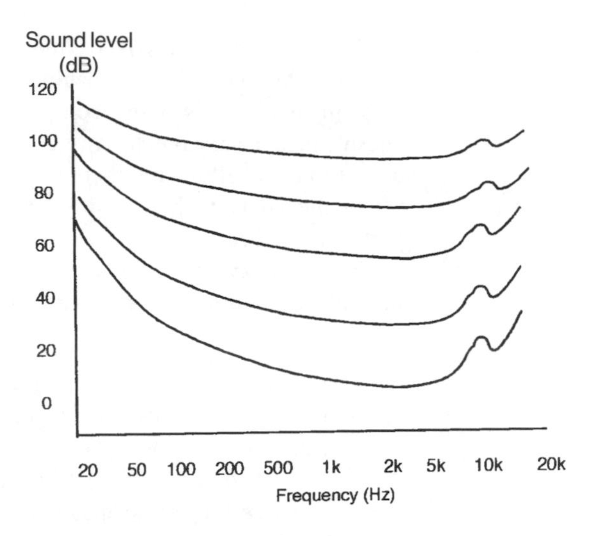

Figure 7.2 Equal loudness curves

people so that they are representative of the response of the average person's ears. These are, remember, for the average person. They will almost certainly not apply to any particular individual. Also illness and exposure to loud noises will be likely to cause differences in the shape of the curves.

The bottom curve is the threshold of hearing and is, as we have seen, far from a straight line. However at higher sound levels, when the sound pressure level is around 80 to 100 dB, the curves are nearer to straight lines. At 20 Hz the ear is about 20 dB less sensitive than it is at 3 kHz, whereas at the threshold of hearing it is about 70 dB less sensitive.

Some knowledge of the implications of equal loudness curves is of considerable importance to anyone working in sound. The point is that the frequency range one hears depends on the level of the sound at the ears. At low listening levels the bass frequencies may not be heard. (The high frequencies may also be lacking but to a much lower extent.) At high levels a much wider range of frequencies can be perceived. Therefore any serious listening, or *monitoring*, should be done at fairly high sound levels. (But see later when we go into the risks of hearing damage!)

The reader can easily investigate the difference in frequency response at different loudnesses by playing a piece of music, preferably with plenty of bass, over a loudspeaker first at a comfortably loud level and then with the volume turned low. The bass will be seem to be much reduced at the low listening level.

Many of the better domestic hi-fi systems have a switch or button marked 'bass control', or something like that. When this control is switched in there is some compensation for the ear's lack of bass response at low sound levels.

7.2 Loudness

You can't measure loudness! Loudness is a totally subjective effect – it's like colour. You and I would agree about calling something green because we've grown up to call grass 'green', but the actual sensations that we receive might be totally different. We'd never know!

To take a very obvious example, a person who has a slight hearing impairment would think that the sound from the television set is too quiet while another person might think it was too loud. And it's no good dismissing this by saying that the first person isn't normal in their hearing and therefore doesn't count. None of us have hearing response curves which are exactly similar, so in a sense we're none of us 'normal'!

It follows from this that there's no such thing as a loudness meter. However it is possible to produce a pretty effective alternative. Figure 7.3 shows a simple circuit consisting of a microphone, and amplifier (to 'boost' the very weak electrical signals from the microphone), and an electrical meter that would show the microphone's output.

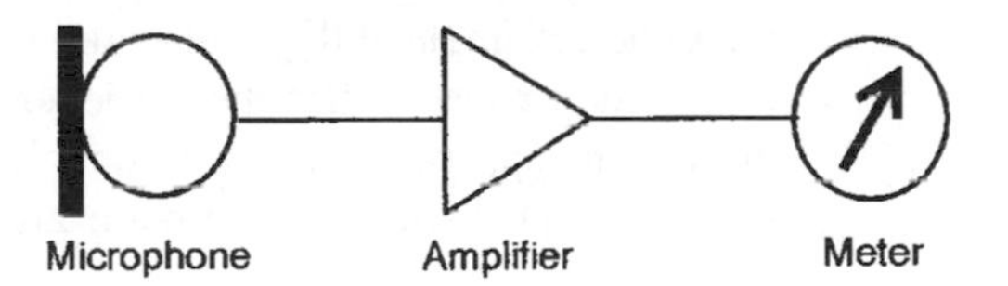

Figure 7.3 A not very good loudness meter!

This device might work reasonably well for high sound levels, where the ear's response is not too far from flat. But it would be hopelessly inadequate at very low sound levels.

The answer, which was adopted very many years ago, is to build into the metering device some electrical circuits (*filters*) which correspond to the ear's response.

Originally there were three filters, designated A, B and C. The idea was that the A filter (or 'network') would correspond to the average ear's response to fairly quiet sounds, B to medium loud sounds and C would have a more-or-less flat response and be appropriate for loud sounds. Consequently sound level meters had a switching system to allow the correct filter to be connected in. After some years of experience, though, it was found, rather surprisingly, that the A filter gave results, *regardless of the level*, which seemed to agree reasonably closely with people's judgements of loudness. Most modern meters, then, have a choice of 'A' or a flat response (which is useful in some circumstances. The flat response setting is sometimes called 'C' or 'LIN' for linear).

Measurements made with a meter, calibrated in dB, and using the A circuit, are designated dB(A), or dBA. Such measurements are often called 'loudness' measurements, which strictly they are not, but they can be a useful approximation to most people's assessments of loudness. The table gives some examples. Note that they are only reasonably typical values of dBA. There can be quite marked variations in, for example, different people's choices of the sound loudness of their television sets.

One point must be emphasized. Human beings are pretty hopeless at judging loudness! We are very much conditioned by circumstances. For example a sound that we find irritating we tend to assess as being louder than it really is – a neighbour's radio might in fact register quite a low number of dBA, but if the radio is getting on our nerves we'd estimate it at a much higher value (unless we were straining to hear what the music was, in which case it would be too quiet!). Even more important is the fact that we are conditioned by noise levels we have recently been exposed to. A good example of this is that after listening to some music for some time we tend to find the music not loud enough and we turn the volume up. An even better example is that a car radio might seem very quiet when driving home after a rock concert!

Description	**Noise**	**dBA**
Painful	Jet engine, near	140
Deafening	Jet aircraft at 150 m	120
	Orchestra, loudest at 5 m	110
	Loud motor horn at 5 m	100

Description	Noise	dBA
Very loud	Small jet aircraft at 150 m	95
	Busy street, workshop	90
	Average orchestra at 5 m	80
Loud	Television set, full volume, at 2 m	70
	Normal conversation at 1 m	60–65
Moderate	Inside fairly quiet car	50
	Inside quiet house	40
Faint	Rural area at night	30
Very faint	Sound-proofed room	10

7.3 Hearing damage

Not surprisingly, after what we have said about its extreme sensitivity, the ear is a delicate instrument and can be damaged. Curiously, perhaps, the part which we might think is the most vulnerable, the eardrum, is relatively rugged. It can be damaged by loud explosions nearby but can repair itself in a few weeks in a person in good health and of not too advanced age. (It should nevertheless never be unnecessarily exposed to any risk. Young children must be discouraged from poking pencils into it. It has been said that you should never put anything smaller than your elbow inside your ear!)

It is the delicate internal mechanism which can be damaged by exposure to excessive noise. Without going into the structure of the ear it must be enough here to say that there is a membrane which vibrates under the influence of sound vibrations. Connected to it are *hair cells*, small hairs attached to nerve endings, and when the membrane is set into movement the hair cells send signals along nerves to the brain. The region in which these hair cells are fixed is particularly delicate. Once damaged by excessive sound it can never be repaired and there is thus hearing impairment in that region.

A great deal of investigation has been carried out on the causes of hearing impairment. (The word *deafness* suggests a total loss of hearing and that in fact is quite rare. Usually there is a reduction in sensitivity which may occur over a limited frequency range. The amount of hearing loss may be quite mild or it may be very severe.) It's all a matter of *sound exposure* – the duration and the sound level both have to be taken into account. It's generally agreed that there is only a small risk to a person's hearing for sound levels below 90 dBA for any length of time. However, above 90 dBA the exposure time becomes reduced, halving for each 3 dBA increase in level. The table shows this.

Sound level in dBA	Permitted exposure time
90	8 hours
93	4 hours
96	2 hours
99	1 hour
102	30 minutes
105	15 minutes
108	8 minutes
111	4 minutes
114	2 minutes
117	1 minute

A big problem is this: the table shows permitted exposure times assuming the noise level is constant. But it rarely is, or even if the noise level doesn't vary then a worker is not likely to be in that noise environment for a very long time. There are such things as meal and coffee breaks.

To get round this the concept of *equivalent noise level* or, *personal noise dose* (the two are nearly the same thing), has been developed. The equivalent noise level is denoted by L_{eq}. If we have a varying noise, as is usually the case, its L_{eq} is the level in dBA which has the same effect on the ear as a steady noise. For example, in a studio the noise level in a control room might be 100 dBA at certain peaks, no more than 80 dBA for a lot of the time, but less than 75 dBA while adjustments are made to instruments and equipment, and less even than that at coffee breaks. A suitable meter, known as an *integrating sound level meter*, can take times and levels into account and could show a reading for the L_{eq} for the time in question.

However in the somewhat simplified case illustrated in Figure 7.4 L_{eq} works out at about 85 dBA, so anyone in that environment for the whole of the 8 hours would be reasonably within the 'safe' limits.

It is a requirement by the UK's Health and Safety Executive that where the L_{eq} is 85 dBA but under 90 dBA ear defenders must be made available. At 90 dBA and above ear defenders are compulsory and noise levels must be reduced where practicable. It has proved difficult to enforce these laws in recording and broadcasting studios.

Before leaving the subject of the risks of hearing damage we

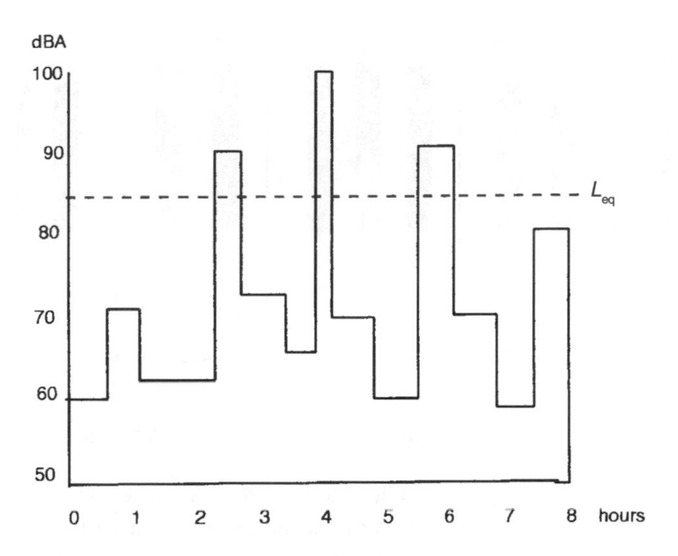

Figure 7.4 The equivalent of a varying noise level

should add that where we have used L_{eq} we should probably have used the symbol for *personal noise dose*, which we mentioned briefly earlier, $L_{ep,d}$. Essentially L_{eq} refers to the equivalent noise level in a work place; $L_{ep,d}$. refers to the noise exposure of an individual.

7.4 Pitch

We all have a pretty good idea of what is meant by the word 'pitch' and yet it isn't easy to put it accurately into words. One definition is 'the position of a note in the musical scale'. But that leaves open the question of what a musical scale is – a range of notes of different pitch! We'll just say at the moment that pitch refers to the 'highness' or 'lowness' of a note. We can, though, go rather further than that and relate pitch to frequency – the two are very closely related. In fact it's convenient to give an exact frequency to a particular musical note, or rather the *fundamental frequency* of that note but we ought to note that the pitch of a sound can be affected by its loudness. However that is getting into rather deep waters, so we'll content ourselves by assuming that pitch can be expressed as a frequency.

Readers with musical knowledge, and in particular those who play any kind of keyboard instrument, will recognize Figure 7.5.

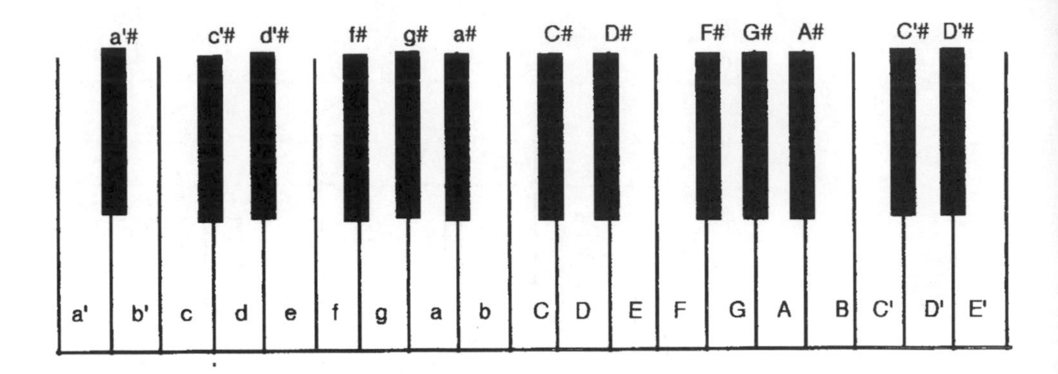

Figure 7.5 Section of a keyboard

The diagram shows a section of a typical keyboard. The letters of the notes repeat a' to a, a to A and so on. Each of these sections covers eight white notes and is called an octave. But if the black notes are counted as well there is a total of twelve. Here is not the place to go into musical scales which can be a very complicated subject and what follows is somewhat simplified – a trained musician might raise his or her eyebrows slightly! We'll just say that the twelve notes are, to the Western ear, equally spaced. They are called semitones. Going from C to C# is one semitone and appears to be the same increase in pitch as going from F to F# which is another semitone. (The # is called a sharp sign and means the note which is a semitone higher. C# is a semitone higher than C, and so on. Although we haven't used it here there is another symbol *b* which means that the note is *flattened* – lowered by a semitone.)

Another very important point, and one that's very difficult to explain in words, is that somehow two notes an octave apart have something in common. It's as though they are the same note but one is higher than the other. Put another way, if a singer were played a note too high or low for their voice they would sing the note an octave (or maybe two octaves) away and it would sound perfectly acceptable.

An octave represents a doubling or halving of the frequency

A most important fact about octaves is that they represent a doubling or halving of the frequency. For example, musicians adopt, as their standard note for tuning, *International A,* which has, by general agreement, a frequency of 440 Hz. A note an octave above this has a frequency of 880 Hz, a note the octave below is 220 Hz.

> A semitone is equivalent to a frequency change of about 6%

Now, if there are twelve equally spaced notes in an octave, and an octave represents a factor of 2 in the frequency, then the frequency change from one note to the next is 12 + 2. If the reasoning behind this is not clear don't worry. Take it as a fact.

The important point is that 12 + 2 is equal to 1.059 or about 1.06. Put another way, a semitone change in pitch is very close to a frequency change of 6%. We are now in a position to draw up a table of notes and their approximate frequencies given at the end of this chapter.

The curious reader might wish to confirm that the frequencies of adjacent notes are close to 6% in their frequency ratios. (They would be closer but for the fact that the frequencies in the table have been given to the nearest whole number.)

7.5 A word of caution

Before we leave the subject of hearing we should emphasize the fact that the ear/brain system is easily fooled! It's sometimes very difficult or even impossible to convince oneself that one isn't being deluded. Stereo is a delusion (or illusion!) as we shall see. What we think we are hearing can be affected by what we see or what we think we *ought* to be hearing. The author once carried out a series of tests with trained and experienced listeners to get them to assess the quality of different loudspeakers. Some of these tests were carried out twice – once when the listeners could see the loudspeakers and once when the loudspeakers were hidden behind very thin gauze curtains. It was amazing how often the assessments were affected by what the listeners could see.

Note	Frequency (Hz)	Note	Frequency (Hz)
a'	110	C	262
a'#	117	C#	277
b'	123	D	294
c	131	D#	311
c#	139	E	330
d	147	F	349

Note	Frequency (Hz)	Note	Frequency (Hz)
d#	156	F#	370
e	165	G	392
f	175	G#	415
f#	185	A	440
g	196	A#	466
g#	208	B	494
a	220	C'	523
a#	233	C'#	554
b	247		

Also it's all too easy to make an adjustment to a control and be well satisfied with the result – only to find later that the adjustment had been made to a control that was out of the circuit. The reader may laugh, but almost all workers in professional audio will have seen something like that happen – never to themselves of course, always to someone else!

8

Basic acoustics

8.1 What do we mean by good acoustics?

We often hear people talking about 'good' or 'bad' acoustics but when one comes to analysing these statements they're often rather vague. Here we're going to say that good acoustics do not impair, or may even improve, the sound, whether for hearing by a live audience or for recording or broadcasting. (It doesn't necessarily follow that good listening conditions mean that the venue is good for using microphones in, and *vice versa*.) Poor acoustics can render speech unintelligible, or intelligible only with difficulty, music can be 'blurred' or sound unnatural, and so on. The subject of acoustics is a complex one: to some extent it is an art and at the same time it is a science. Also it often involves trial and error.

In this book we can give no more than an outline of the subject.

Terminology: we shall use the term 'room acoustics' to include concert halls, radio studios, television studios, in fact any venue in which microphones might be placed.

8.2 Avoiding resonances

In an earlier chapter we outlined the behaviour of sound waves in pipes which were open at one end or open at both ends. We also raised the question of a pipe which was closed at both ends, and added that this wasn't as silly as it seemed because it had relevance to rooms.

As a reminder, when there is a resonance between the two *open* ends of a pipe there is a pressure node at the ends, because the pressure must be the same as the outside air.

If we have two parallel surfaces – or walls in the case of a room – there must be pressure *antinodes* at the walls. Figure 8.1 shows this. The uppermost diagram shows the simplest possible mode of resonance, in which there is one half wavelength between the walls. In the middle diagram there is one wavelength and in the lower one we have one and a half wavelengths. If the wall spacing is l then there are possible resonances when l is equal to $\lambda/2$, λ, $3\lambda/2$ and so on. A simple formula for the frequencies of resonance makes use of $c = f\lambda$. If we rewrite that as $\lambda = c/f$ and substitute for λ we can find a general rule for resonant frequencies:

$$f = nc/2l,$$

where n is 1,2,3, etc.

Figure 8.1 Resonances between two parallel walls

Now before we go any further let's see what the significance of this is. Resonances of this kind are the result of two waves going in opposite directions and that means there are places where they cancel each other – producing nodes – and other places where they reinforce each other, and produce antinodes, terms which we explained in an earlier chapter

If we imagine a microphone placed somewhere – anywhere you like – between the walls shown in Figure 8.1 then it's going to be at a node for some frequencies, an antinode for some other frequencies, and part-way between a node and an antinode for yet other frequencies. And since all microphones are operated by sound wave pressure it means that if we've strong acoustic resonances present any microphone cannot possibly pick up a correct version of the original sound. Some frequencies will be exaggerated, others will be at reduced level, or possibly even missing totally.

Such resonances are called *standing waves*, which is a bit misleading as the waves are not actually standing. The nodes and antinodes have fixed positions, but that's all. However the term is the standard one so we will use it. The point is that standing waves in a room or, more particularly, a studio are clearly undesirable.

The reader may perhaps wonder why standing waves are not more obvious in everyday life. There are three good reasons. One is that if standing waves are present in one's sitting room, for example, one of our ears may be at a node and the other may well not be. Secondly, our brains are well adapted to correct for and ignore to some extent information reaching the ears which is not important, and the third reason is that with ordinary speech and music the pattern of nodes and antinodes is changing very rapidly and while a microphone would be picking up an obviously incorrect version of the original sound it would not be easy to identify the individual nodes and antinodes. However, a fairly simple experiment can demonstrate the existence of standing waves in almost any room. All that is needed is a medium loud steady pure tone – between about 500 Hz to 2 kHz should work well. If that can be produced then all the experimenter needs to do is to stop up one ear firmly with the hand and then walk about the room. The loudness of the tone will change with one's position.

All that we've said so far is a simplification of the facts. Any normal room has two pairs of facing walls, a floor and a ceiling. There are then *three* pairs of facing surfaces, and three sets of standing waves, all with their frequencies determined by the spacing between the surfaces. And as if that weren't enough there can also be standing waves between facing edges where two flat surfaces meet and even between opposite corners of a room! The pattern of standing waves can be very complicated. Figure 8.2 shows the directions of the three standing waves between the walls floor and ceiling, one edge to edge and one corner to corner.

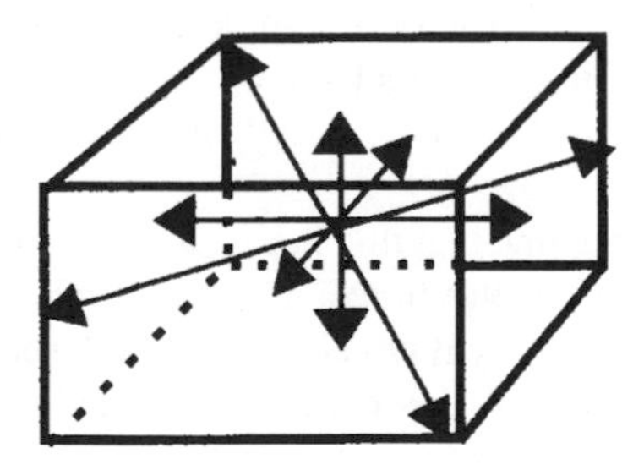

Figure 8.2 Standing wave modes in a room

Next, how to avoid standing waves, or at least reduce them to acceptable levels. For a start any sound absorbing materials are obviously going to reduce the amount of reflected wave at a surface. We shall go into sound absorbers later in this chapter but we can say here that massive amounts of absorber would kill all standing waves. This would be a quite impracticable method (a) on grounds of cost and (b) because, as we shall see, there are likely to be undesirable effects if all reflections are removed. The best, easiest and probably the cheapest way of reducing standing waves is to provide *diffusion*. This means having irregularities which break up the reflected waves.

There are many ways of providing diffusion. One is to use more-or-less carefully designed *diffusers*. Some concert halls have suspended structures which act in this way. In television studios the sets and technical equipment can generally be relied upon to provide adequate diffusion; some radio and recording studios have walls and ceilings with devices which stand out from the surfaces and act as diffusers and sound absorbers at the same time. Also there are rather sophisticated units with random depths providing diffusion which is effective over a wide frequency range.

8.3 Reverberation

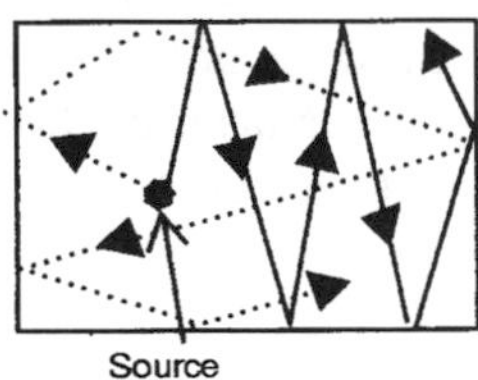

If a sound is created in a room it travels out until it hits walls, floor, ceiling and other surfaces. The likelihood is that most of the sound energy striking a surface will be reflected away, following incidentally the same laws of reflection as light: the angle of incidence is the same as the angle of reflection. The reflected sound waves then go on to strike other surfaces, again much is reflected and so on, until the sound wave energy is too weak to be detected. This process is called *reverberation* and is illustrated for two sample sound wave paths in the margin diagram. Graphically the decay of the sound pressure follows what mathematicians call an *exponential* curve and is shown in Figure 8.3.

The time that this decay takes is very important. But, how do we define the time an exponential curve takes to die away? Theoretically it never gets to zero! The answer was given by an American by the name of W.C. Sabine who was the first person to carry out serious scientific experiments on reverberation. We

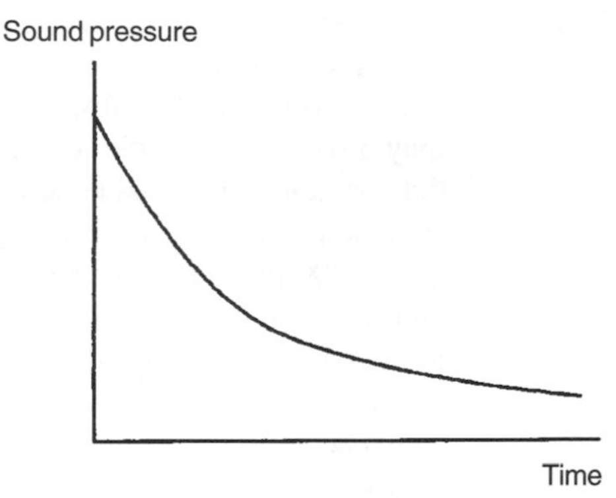

Figure 8.3 The decay of sound pressure in a room

Reverberation time (r.t.) is the time it takes for the reverberant sound to decay through 60 dB

define *reverberation time* as the time it takes the reverberant sound to decay through 60 dB. (Actually decibels didn't exist in Sabine's time – the late 1800s – and he worked on the time it took for the sound intensity to die away to one millionth of its original, i.e.. 10 log 1 000 000 = 10 x 6 = 60.)

Figure 8.4 illustrates reverberation time graphically. Note that if we use a decibel vertical axis the line of the decay becomes a straight line. At least it does in a rather idealized picture. In reality the line would only be reasonably straight and would almost certainly be wiggly at its lower end.

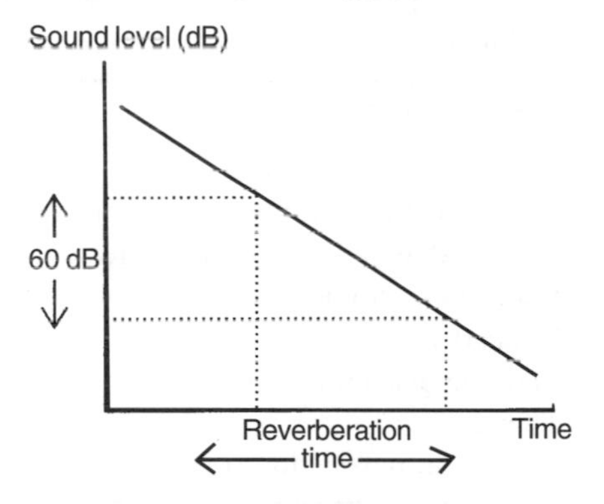

Figure 8.4 Reverberation time shown graphically

Reverberation time, often abbreviated to R.T., r.t. or T60 (we shall use r.t.) has a great effect on the acoustics of a studio or concert hall. Furthermore the r.t. which is appropriate to an environment depends on the use, and in the case of a studio, whether it is for radio or for television. If it's for radio, then what type of programme? We can perhaps explain this by a few examples:

1. In a talks studio for radio if there is too long an r.t. then the reproduced speech sounds 'echoey', or remote. If it is too short then the effect is somewhat unnatural (and may also be very unpleasant for the speaker(s)). About 0.4 s is generally considered to be about right.
2. A moderate-sized conference room, seating perhaps about 200 to 300 people, needs an r.t. of around 1 s. Much less and the sound will be absorbed too quickly and people at the back may have difficulty in hearing, unless there is a loudspeaker system. Much more than 1 s and the sound becomes indistinct because spoken words overlap, as it were.
3. A rock music studio should have a short r.t., perhaps about 0.5 s or less (which may not seem very short but because such a studio will have to be moderately large it might be very difficult to achieve a shorter r.t., for reasons which we'll see later.)
4. Orchestral music studios for recording or broadcasting will generally need r.t.s of, if possible, 1.5 s or even more. Most concert halls have r.t.s of 2 s or so, and to produce a realistic sound in the studio involves having acoustics which are as similar as possible to the real thing. (Costs and other economic matters mean that the acoustics of studios are often a compromise.)

The table below lists some typical values of r.t.

Site or function	r.t. (seconds)
Average sitting room	0.5
Small church with plastered walls, etc.	1.0
Large cathedral, bass frequencies	10–15
Radio talks studio	0.4
Concert hall	around 2
Large television studio	0.7–1.1
Theatre	1
Orchestral music studio for radio	1.5–1.8

The next question is how to obtain as closely as one can the desired values of r.t. How, for example, do architects achieve a value of 0.4 s for a radio talks studio? The answer is: not very easily! One difficulty is that of obtaining the correct r.t. at all frequencies. There is a tradition in most of the UK, but not necessarily elsewhere, that the r.t. should be reasonably constant over the audio frequency range. For reasons which will shortly emerge this can be difficult.

First we must see the factors that affect r.t. Primarily there are two:

1. The volume of the 'room'.
2. The amount of sound absorbing material present.

We will take these in order. First the volume. It was found by Sabine, whom we mentioned earlier, that to a very good approximation:

$$\text{r.t.} \propto \text{volume}$$

($\propto$ means 'is proportional to').

This is not very surprising when one thinks about it. To illustrate this look at Figure 8.5. This shows two rooms with different sizes. For simplicity only the plan views are shown.

In each room let us suppose the sound dies away, through 60 dB, after 5 reflections. Because of the longer paths this is going to take much longer in the large room than in the small one. This doesn't of course prove that it is the volume that the r.t. is proportional to, but it does show that the larger the size of the room the longer the r.t.

The second factor, the amount of sound absorption present, is fairly obvious. If a room had all its surfaces coated with material that absorbed all the sound striking it then the r.t. should be zero. We can, however, look at absorption in a little more detail.

All materials absorb sound to some extent. Hard, shiny materials like plate glass absorb very little – only perhaps one or two percent, and some of that may be a result of the glass vibrating slightly. Others, porous substances in particular, can absorb sound waves very markedly. These substances, and most practical sound

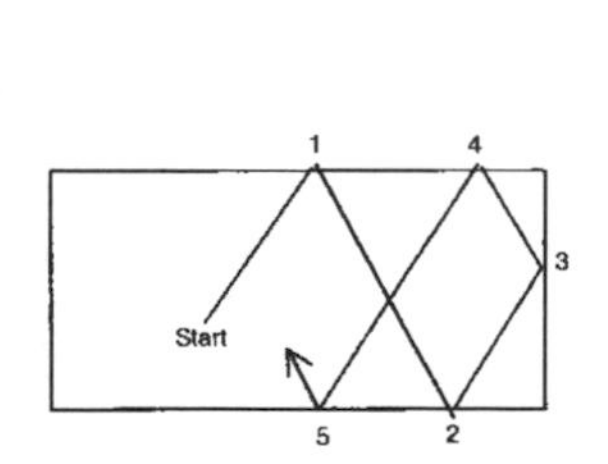

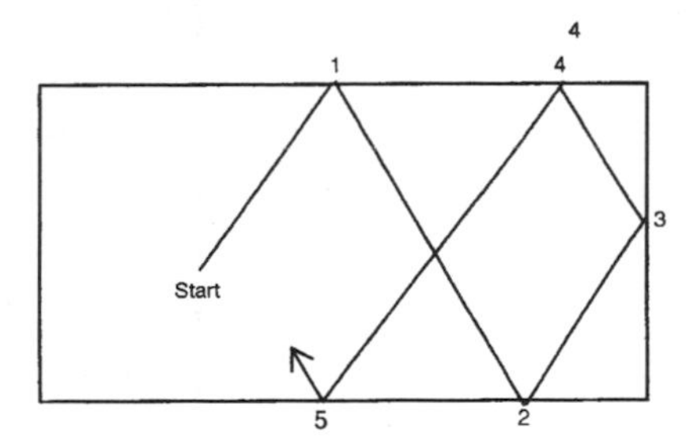

Figure 8.5 Typical sound paths in different size rooms

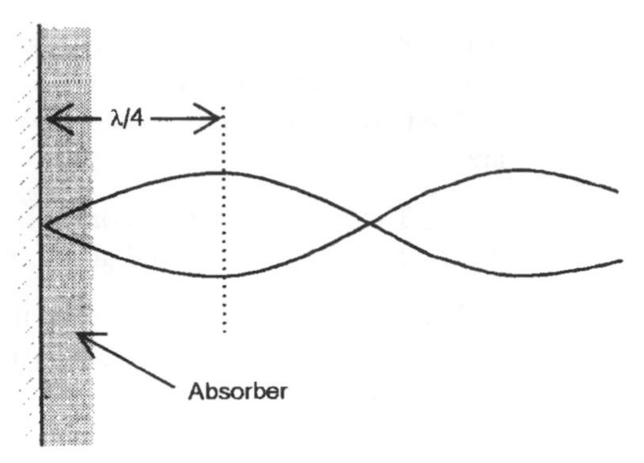

Figure 8.6 Reflection with a small thickness of absorber

absorbers, work by causing sound waves to lose their energy by friction. Figure 8.6 shows a sound wave undergoing reflection at a wall when there is a small thickness of absorber.

Note that the sound wave is shown here not as a pressure graph but as a *particle displacement* graph. At the wall there is no movement of particles because they cannot burrow into the wall so there is a *displacement node*. At a quarter of a wavelength away there is a displacement antinode, a region where there is maximum particle displacement. In Figure 8.6 the absorption will not be very great because the absorber isn't thick enough to catch the maximum displacement. Figure 8.7 on the other hand shows a much more effective situation where the absorber extends out to at least a quarter of a wavelength.

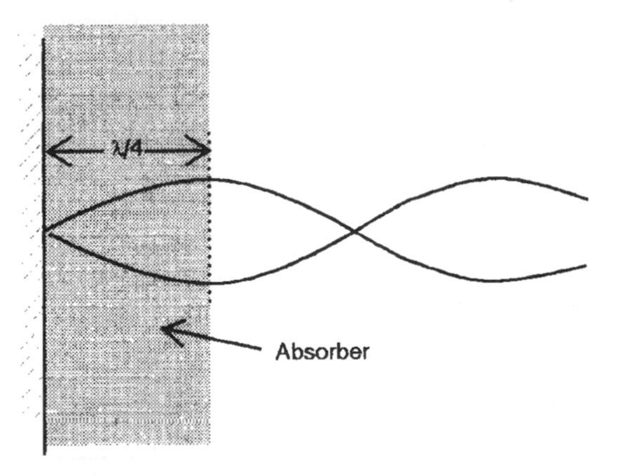

Figure 8.7 Reflection with a greater thickness of absorber

From this we can see that a porous absorber should be a quarter of a wavelength thick to be really effective. This presents a problem because, if we want to absorb frequencies of 30 Hz, which is currently about the lowest useful frequency that is broadcast or recorded, the corresponding wavelength is about 10 m, so we should need absorbers around 2.5 m thick! This isn't practical but, as we shall see, there are alternative answers.

At this point we should introduce the idea of the *absorption coefficient*. This is usually denoted by the Greek letter α (alpha). α is defined as the fraction of sound which is absorbed by a material.

Thus if a material absorbs *all* the sound which strikes it the value of α is 1. If it absorbs no sound (if the sound is reflected away, or diffracted round it), then α = 0. (Sometimes absorption coefficients are given as percentages. A perfect absorber would have a coefficient of 100%.) The table below gives a few typical absorption coefficients. Note that the values of α are given at three different frequencies.

The reader will notice that in each case there is a tendency for α to increase as the frequency gets higher (breeze block absorption drops off a little for some reason but even so the absorption is greater than at low frequencies) and this should be expected after what we were saying about the action of porous materials.

Material	**Absorption coefficient, α, at**		
	125 Hz	**500 Hz**	**4000 Hz**
Brick wall	0.02	0.03	0.07
Unplastered breeze blocks	0.25	0.60	0.45
Heavy drape curtains	0.1	0.4	0.5
10 mm thick carpet	0.09	0.21	0.37

We can now begin to see how the room volume and the amount of absorption affect r.t. First though, we need to see what is meant by a *unit of absorption*. This is the equivalent of one square metre of perfect absorber (α = 1). The unit is given the name of the *sabine* (pronounced 'saybine'). Sometimes the last letter is dropped, making it *sabin*.

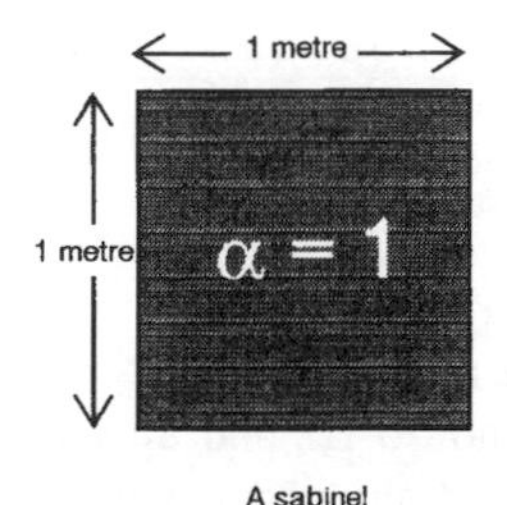

A sabine!

The numer of sabines present in a surface, such as a wall, is found by multiplying the absorption coefficient by the area of the surface:

number of sabines = $A \times \alpha$

A being the area. This brings us to what is known as *Sabine's formula:*

$$\text{r.t.} = \frac{0.16 \text{ x volume (cubic metres)}}{\text{total number of sabines present}} \text{ seconds}$$

The '0.16' is simply a constant that gives an answer in seconds when the dimensions are in metres. It is different – 0.05 – if measurements are in feet.

We should add that Sabine's formula as we have given it is an approximation. It's reasonably accurate when there is relatively little absorption but fails when absorption is large. The reader might care to calculate the r.t. for a room of any size when all the surfaces are covered with 100% absorber. The r.t. ought to be zero, but it will be found that the r.t. calculated in this way isn't zero! There are, though, more accurate versions of the formula.

So now we can see that for a given volume of studio to have the wanted reverberation time it's a matter of putting in the correct number of sabines or, to be more practical, the correct amount of absorber whose absorption coefficient is known. This is easier said than done because, as we've seen, most porous absorbers work better at the higher frequencies, so there has to be a certain amount of juggling of kinds of absorber (and trial and error, very often) to achieve not only the wanted r.t. but also to have that wanted r.t. over a wide frequency range.

There are other types of specialized absorbers than purely porous ones. We will mention just one here, the *wide band porous absorber*. Its construction is shown in Figure 8.8.

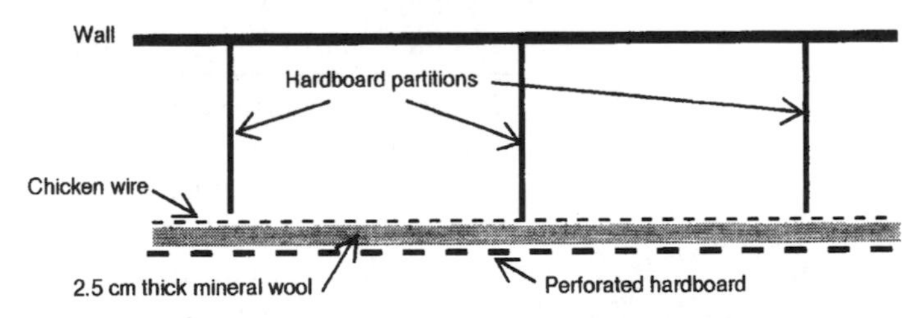

Figure 8.8 Wide band porous absorber

Wide band porous absorbers can have a high value of α over a wide range of frequencies – from around 70 Hz and upwards. Also they can be made in modular units of any convenient size, 600 mm square is common, so they can be fitted to walls in patterns which are effective acoustically and also reasonably pleasing visually. Their depth can vary but is typically some 200 mm, which is enough to provide some diffusion, a point we made earlier.

Television studios present a special problem. To begin with a major production studio is large: 30 m or so long, 25 m wide and 15 m or more high. The difficulty, in acoustic terms, is that there have to be large areas of non-absorbent surface. The floor, for example, has to be hard, and there will also be signif-icant areas of panels for plugging in cameras, microphones and similar essentials. The amount of surface available for acoustic treatment is therefore restricted, and the consequence is that a large studio cannot have a short r.t. In the table of typical r.t.s we gave television studios as having values of 0.7 to 1.1 s, and that sort of thing is about the least that can be managed. This makes television sound something of a problem (well, it can be a problem for very many reasons, and this is just one!). To begin with, the r.t.s that are achievable are too long for speech, too long for rock music, too short for orchestral music, and that's only part of the story. The answer has to be found in things like good microphone techniques. The use of external venues like concert halls for orchestral music is one way of getting round the difficulty.

8.4 Sound isolation

In our somewhat simplified account of acoustics we are going to define sound isolation as keeping external noises out of studios. We might add that in some cases, for instance radio and recording studios used for rock music, the problem may well be that of keeping the noise in! It could seem at first sight that we would simply want to sound-proof all studios so that no external sounds could penetrate. That, though, is actually very unrealistic. For a start complete sound proofing is very expensive. For example, reinforced concrete 1 m thick will have a sound reduction index, as it's called, of 45 to 50 dB at 125 Hz and 65 to 70 dB at 1 kHz. In other words, to have 0 dBA noise level in a studio with walls of this thickness concrete, and a roof/ceiling of similar acoustic properties, the maximum noise outside must be no more than about 50 dBA at 125 Hz and 70 dBA at 1 kHz. But ordinary speech is about 65 dBA!

So it has to be recognized that perfect sound insulation is not possible, and in any case, all studios have a level of internal noise, so there's no point in trying to achieve lower levels than that.

Studios for radio drama are the most critical: there can be

'dramatic silences' in which the slightest intrusive sound could destroy the effect. Imagine the dialogue of a Shakespearean play with a police siren audible in the pauses!

Television studios are less critical, for two reasons: first there is always some background noise in the studio caused by things like air-conditioning and the movement of performers and crew, and secondly the fact that the 'consumer' at home has the pictures there can be some slight relaxation of the standards of the sound.

So what we are saying is that designers have certain criteria to work to in deciding what kind of sound insulation to provide for a given studio. And, incidentally, a lot can be done with common-sense approaches, like siting studios away from major noise sources, having offices or scenery stores on the outside of the studios to provide a kind of shield from traffic and industrial noise, and so on.

Quite apart from reducing the effects of external noise there is the matter of isolating internal sounds. A very important example is with a studio and its associated control rooms, in particular the sound control area. The latter will have loudspeakers at possibly a fairly high volume, and if this sound gets back into the studio and is picked up by the microphones there may easily be an undesirable colouration of the sound output. Thus, all doors need to have special seals, there are very often two sets of doors providing what is sometimes called a 'sound lock' and observation windows have to be double- or triple-glazed. On this point it is worth mentioning that for double glazing to be acoustically effective the spacing needs to be very much greater than the few millimetres used in domestic thermal double glazing. 10 cm is a bare minimum and 20 cm is more common in recording and broadcasting.

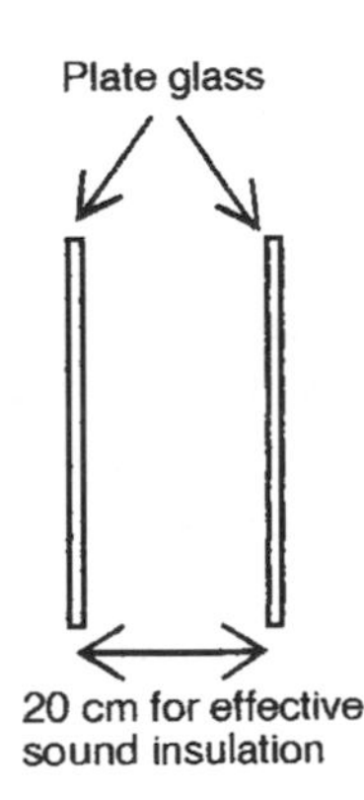

And finally, it's vital that all gaps are sealed. Sound waves can pass through really quite small crevices, so places where cables and other services enter the studio must be treated with suitable materials such as bitumastic compounds.

To summarize we can say that acoustics which are ideal hardly ever exist. If they are good for an audience they may be somewhat unsatisfactory for broadcasters and recording companies. And pleasing the performers is far from easy. So one has to accept that it's an imperfect world in this respect! It makes intelligent use of microphones and other technologies very important. But we'll come to those later.

9

Microphones

9.1 The basic parts of a microphone

Everyone knows what a microphone is – 'a device which converts sound waves into electricity' would be a typical definition. It's not, actually, a very accurate one because it's not necessarily true to say that the sound waves are converted into anything. A better definition says that a microphone produces an electrical voltage which is, as far as possible, a *replica* of the sound waves which strike the microphone, or part of it.

A knowledge of the behaviour of microphones is essential for anyone working in sound and in this chapter we shall try to explain their behaviour. We'd add that some understanding of how microphones work can be a very important part of knowing how they behave.

Basically a microphone can be thought of as consisting of three components:

1. A part which is made to vibrate when sound waves strike it. This is called the *diaphragm*. It can take a variety of forms but by far the most common is a circular disk of very thin plastic or metal, or a combination of both, and while the size can vary a typical diameter is in the region of a centimetre. It can be much less – only a very few millimetres – and large ones can be between two and three centimetres across.

The transducer produces an electrical signal which is a replica of the sound waves.

2. Something which creates a small electrical voltage which ideally follows exactly the pattern of sound wave pressures hitting the diaphragm. The term for this part is transducer. (In more general terms a transducer is any device which converts one type of energy or motion into another. As we shall see, loudspeakers have transducers which convert electrical energy into movement to create sound waves.)
3. The microphone case. This might seem surprising and the reader new to the subject could be forgiven for thinking

that the only function of the case was to protect the workings. Later in this chapter we shall see how the nature of the casing affects the way in which sound waves reach the diaphragm and this in turn decides the way in which the microphone responds to sound from different directions.

Now, although we have given the diaphragm as the first item in the list above we shall in fact deal initially with the transducer because the two are usually inseparable and in many microphones the diaphragm is actually part of the transducer.

9.2 Transducers

There are many ways in which movement can be used to create an electrical voltage. Here we will list only the three that are of any practical use in the professional world.

1. Moving coil

Figure 9.1 shows the basic construction.

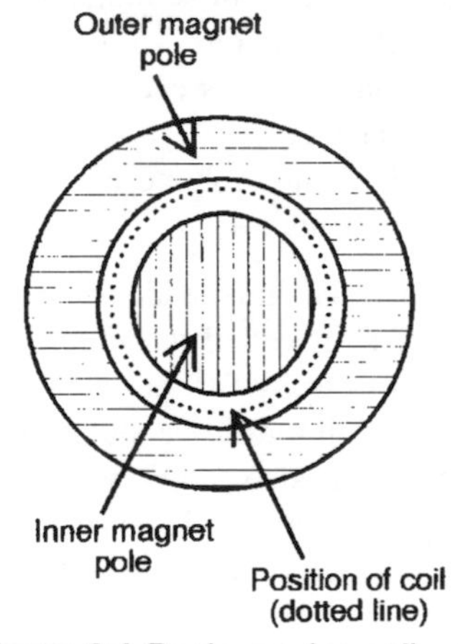

Figure 9.1 Basic moving coil microphone

The diaphragm, typically of very thin aluminium, is domed to make it rigid in the centre and it has corrugations round the outside which allow it to move to and fro (or up and down in the diagram) when sound waves meet it. Attached to it is a coil of very fine wire. There may be about 20 to 30 turns of wire, which is often made of aluminium for the sake of lightness, and this lies in the field of a powerful magnet. Figure 9.1 is a section through the magnet – in fact it's circular with one pole inside the other, as our small diagram shows. And then, not shown, are fine wires connected to the ends of the coil which go off to the microphone's output plug or socket.

The small movements of the coil in the magnetic field generate an e.m.f. This obviously depends upon the amount and velocity of coil movement, which in turn depends upon the pressure of the sound waves striking the diaphragm. It is, however, very small. Taking normal speech from a person about half a metre from the microphone as a rough standard then the output voltage is likely to be in the region of 1 mV. We shall give a table of comparative sensitivities of the three different transducer types later in this chapter.

Moving coil microphones are sometimes called 'dynamic' microphones.

2. Ribbon

This, like the moving coil transducer, depends on the movement of a conductor in a magnetic field to generate an e.m.f. Figure 9.2 shows the idea.

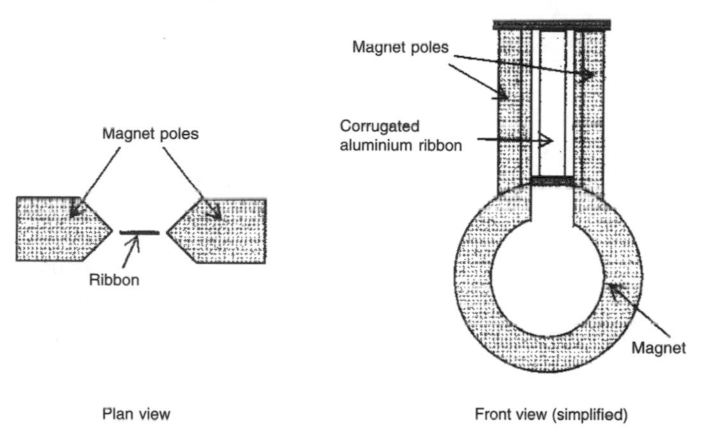

Figure 9.2 Basic ribbon microphone

The strip of aluminium ribbon is simultaneously the diaphragm and the conductor which moves in the magnetic field thus generating an e.m.f. In a typical ribbon microphone the ribbon itself is 2–3 cm long and about half a centimetre wide. It is extremely thin – one famous ribbon microphone which is still to be seen occasionally in broadcasting studios has a ribbon thickness of 0.00006 cm, which is comparable with the wavelength of light! The output voltage of a ribbon microphone, using the same sound source that we used for moving coil transducers, is perhaps in the region of 0.1 mV.

Ribbon microphones are not widely used these days. One reason is that they tend to be rather large and, certainly for television, visually obtrusive. Also, as we shall see, their directional properties, which can occasionally be very valuable, are not as universally useful as some other kinds. On the credit side their ribbon, which, don't forget, is also the diaphragm, is so light that it responds well to the important starting transients of sounds which we referred to in an earlier chapter.

3. Electrostatic transducers

These, sometimes called condenser or capacitor microphones, are very widely used. The principle is shown in Figure 9.3.

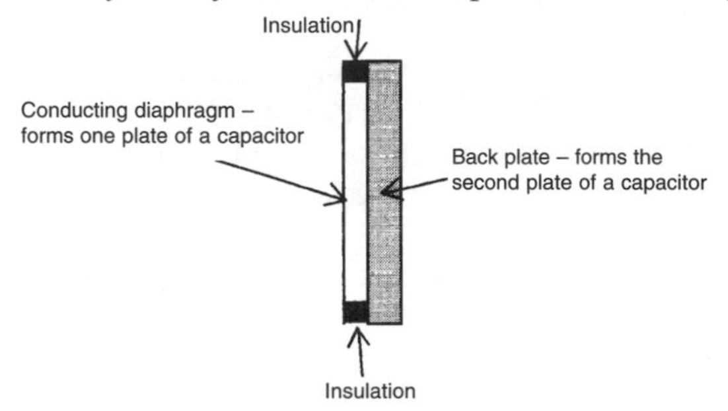

Figure 9.3 Basic electrostatic microphone

The diaphragm forms one plate of a capacitor, the back plate forming the other. The theory is this: if the capacitor carries a charge then there is an electrical potential (voltage) between the plates. The charge Q, the potential V and the capacitance C are related by the formula:

$$Q = CV$$

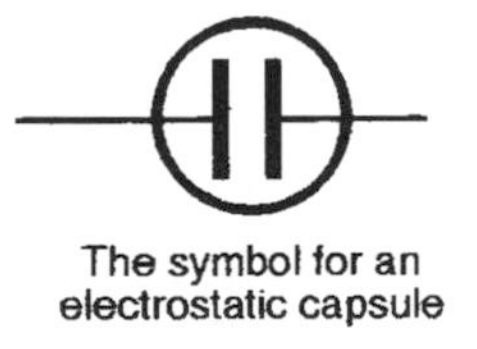

The symbol for an electrostatic capsule

Now if the capacitance C varies, as it will do if sound waves move the diaphragm, then since the charge cannot alter the potential V must vary. It then becomes a matter of taking these voltage variations off to the microphone's output. This has to be done via a small amplifier, usually called the *pre-amplifier* or '*pre-amp*' for short. The reason for using an amplifier is that the *impedance* of the capacitor arrangement, or *capsule* assembly, is extremely high and this renders its wiring very susceptible to interference from external sources – hum from the supply mains for instance. The function of the pre-amp is not so much to raise the level of the signal, although it may well do this, as to bring the effective impedance down to a low value.

There are two ways of giving the capacitor a charge. The original method was to have a source of d.c. with a voltage in the region of 50 to 100 V, as in Figure 9.4. The high resistance *R* is there to 'hold' the charge on the capacitor.

A more recent system which is widely used is to make use of certain materials called *electrets* by analogy with *mag*nets. These are substances which can be given a permanent electric charge – they are subjected to an electric field of many thousands of volts when they have been raised to a high temperature, and then allowed to cool with the electric field still in place. Electret microphones thus do away with the need for a

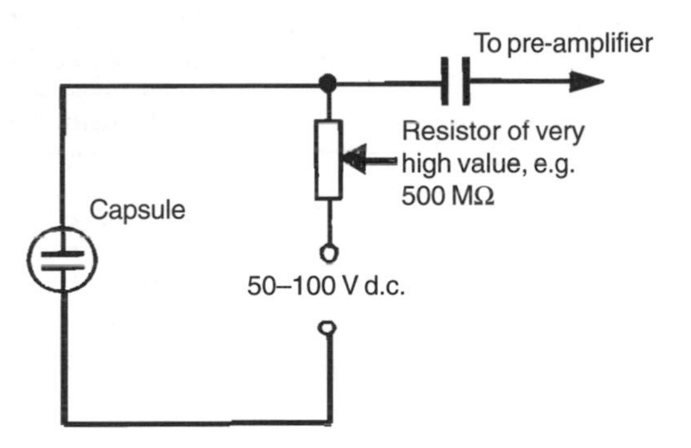

Figure 9.4 A d.c. supply for an electrostatic microphone

Of course there is still the pre-amp which must be supplied with electrical power. In some electrostatic microphones a small battery inside the case is used, but this is inconvenient in a studio using large numbers of microphones. There are, though, ways of supplying the necessary power down the microphone cable – a subject we will deal with later.

There is a specialized type of electrostatic microphone which, although expensive, avoids a major problem. The trouble is that because electrostatic microphones rely upon first-class electrical insulation anything which affects that is undesirable. The worst offender is damp. A microphone can suffer from this if, for example, it is brought from a cold environment into a warm one, when condensation is likely to occur. The effect is a sizzling or crackling noise in the output. It is usually temporary, lasting perhaps half an hour, depending on conditions, but makes the microphone useless until it has dried out.

Microphones which are never taken out of a studio are less likely to be affected in this way. However microphones in mobiles and outside broadcasts (OBs) are much more prone to the problem, and even more so are those used by, for example, film crews doing documentaries and the like.

There is a type of microphone which gets round this problem by using the capacitor principle in a different way – the capsule is part of a tuning circuit in what is in effect an f.m. receiver. The latter is fed with an a.c. signal of fixed frequency (around 8 MHz) so that as the capsule capacitance varies, so does the output of the 'receiver' – properly called a *discriminator*. This system is almost entirely unaffected by humidity problems and therefore is widely used in microphones which have to be used out-of-doors, particularly of the 'gun' type (which we shall deal with later). These microphones are known as *r.f. electrostatic*.

Comparison of transducers

		Transducer	
	Moving coil	*ribbon*	*Electrostatic*
Sensitivity	Medium	Low	High
Robustness	Good	Poor	Good
Sound quality	Good	Can be Excellent	Very good or excellent
Affected by:			
Moisture	No	No	Yes, unless r.f. type
Movement	Generally good	Poor	Generally good
Reliability	Good	Good	Good
Cost	Rather high	High	Quality for quality, often cheaper than the other two

9.3 The important responses of a microphone

It is utterly obvious to say that a professional microphone should produce a good quality output, meaning that it should give a faithful reproduction of the sound waves striking the diaphragm, but how do we measure this? Briefly, not easily. To a large extent the only way to compare the quality of one microphone with another is by listening tests, of which more later. For example we've already emphasized the importance of starting transients, but there's no easy way of assessing a microphone's ability to reproduce them using measuring devices. Carefully carried out

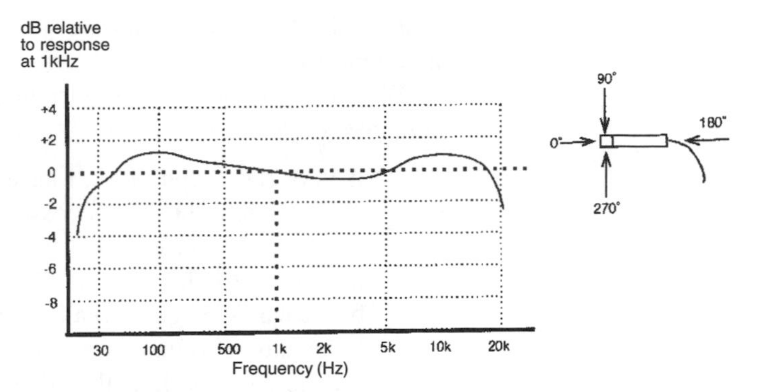

Figure 9.5 A typical frequency response graph

listening tests can, though, make a comparison between two different microphones. Despite this there is one set of measurements which, while not fully conclusive, can nevertheless be a most useful indication of a microphone's performance – the *frequency response*.

By this we mean the ability of the microphone to respond to different frequencies and the best method of showing it is by a graph. Figure 9.5 shows a typical frequency response graph.

Note that the frequency range is from about 30 Hz to 20 kHz. We've said that the normal person's hearing stretches from about 16 Hz to 16 kHz, so why is this different? Well, to begin with there is, at the present time, little point in trying to record or broadcast frequencies below 30 Hz. Until recently the technology has not been capable of handling frequencies below 30 Hz (although with digital recording methods this has changed). It is not easy, however, to make loudspeakers of manageable size emit these frequencies satisfactorily, and in any case there are not many musical sounds of great importance in that range. (Church organ enthusiasts might not agree with that statement!)

At the top end of the range 20 kHz is likely to be detectable by young people, and not only that, it is quite easy these days to record and reproduce such sounds.

The vertical scale is in decibels and it is common to take the zero as representing the response of the microphone at 1 kHz as we have done here. A good microphone will have a frequency response which is within 1 dB from perhaps 100 Hz to 12 kHz and within 2 dB outside that. (Beware of microphone manufacturers who show frequency response graphs of their products with the decibel scale compressed so that the graph looks like an almost straight line! But none of the reputable makers do that. Their data are almost always very reliable and honest.)

We should make two important points about frequency response graphs:

1. A very flat graph doesn't necessarily mean a very high quality microphone. It's possible for a microphone with, for example, an excellent transient response but a less than perfectly flat graph still to give good results as judged by the ear, which is after all the final judge. A flat response graph indicates that the microphone is probably good but it is no more definite than that.
2. For some purposes a non-flat response is desirable. A good instance is with microphones for public address speech. If there is a 'boominess' inside a hall or similar auditorium, and this is quite common, then a microphone with a reduced low frequency response can be more useful than one with a flat response.

The second important characteristic is the *polar diagram* or *directivity pattern*, in other words the way in which the microphone responds to sounds from different directions. In fact, for the user of microphones the polar diagram (we shall use that term as being simpler than 'directivity pattern') is often of much more use than knowledge of the transducer, or detailed information about the frequency response. Figure 9.6 shows a simplified polar diagram. The microphone is imagined to be at the centre and the distance out to the graph shows the sensitivity of the microphone at that angle. The sensitivity is usually given in decibels, relative to the sensitivity at 0° – the *on-axis sensitivity*, although voltages are sometimes used.

Notice also that it is usual to put a little drawing of the microphone to help identify the angles in the diagram.

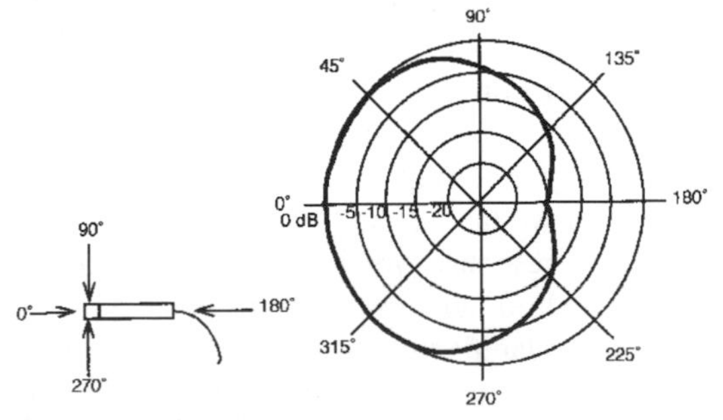

Figure 9.6 A polar diagram

There are five important polar diagrams which we list below:

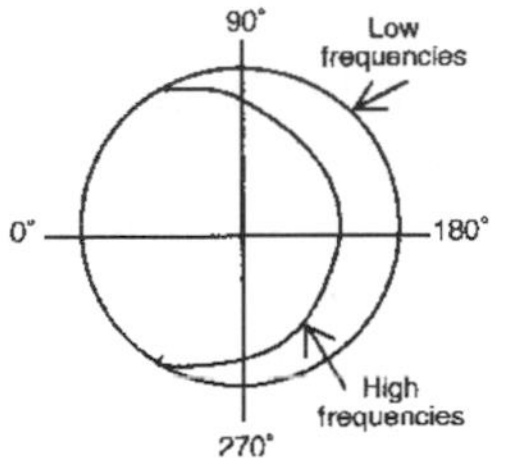

90° Low frequencies 0° 180° High frequencies 270°

1. *Omnidirectional* in which the microphone responds equally to sounds from all directions, as the name suggests. The polar diagram is a circle, at least ideally. Rather surprisingly the omnidirectional characteristic is obtained by allowing sound waves to reach the front of the diaphragm only (as will be seen with other polar diagrams some sound is allowed to reach *both* sides of the diaphragm). The sound waves can do this because they can diffract round on to the diaphragm even when they approach from behind the microphone – provided their wavelength is smaller than the microphone's diameter. However, at the high audio frequencies, with wavelengths of only a centimetre or two, the diffraction process may not occur fully, so the pattern ceases to be a perfect circle.
2. *Figure-of-eight* polar diagram, which is a pretty obvious title as our little drawing shows. This is sometimes called a bi-directional pattern. It's achieved by letting sound waves reach both sides of the diaphragm. The full explanation of how it is obtained is in fact rather complex. It relies on the difference in instantaneous sound pressures on the two sides of the diaphragm – what is properly called the *pressure gradient*. (Figure-of-eight microphones are often referred to as *pressure gradient microphones*.) The phase difference, which may be very small, especially at low frequencies, between the wave at the diaphragm front compared with that at the rear, is enough to cause a resultant pressure difference and thus a force which tends to make the diaphragm move. Ribbon microphones are naturally figure-of-eight microphones, as will be seen from the diagram earlier in this chapter. However electrostatic microphones can also be made to have a figure-of-eight response. This will be explained later.

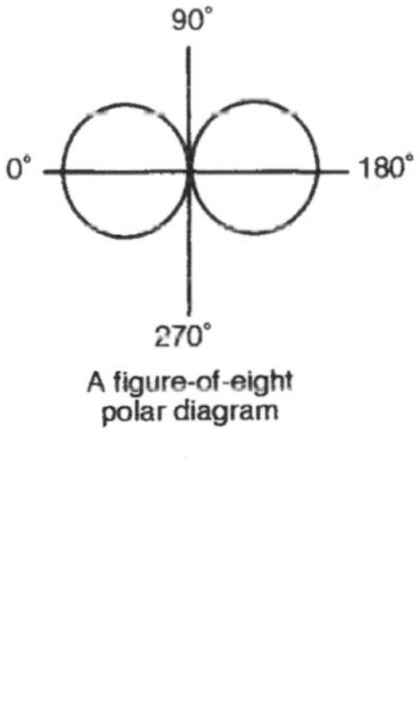

90° 0° 180° 270°

A figure-of-eight polar diagram

   There are certain operational characteristics of all figure-of-eight microphones which the user needs to be aware of. One is that they exhibit an effect known as bass tip-up, an alternative term being proximity effect. What this means is that the bass output is increased when the sound source is close to the microphone, typically half a metre or less. This can be turned to good effect sometimes.
3. *Cardioid*, which means 'heart-shaped' – an obvious name when one looks at the shape. A cardioid pattern is really a

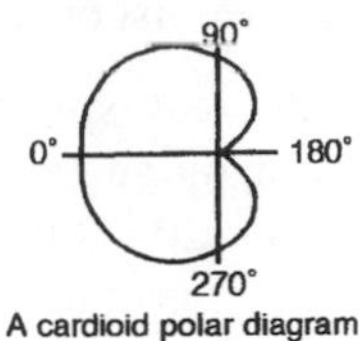

A cardioid polar diagram

combination of an omnidirectional and a figure-of-eight pattern, and indeed at least one early cardioid microphone consisted of a moving coil omnidirectional microphone and a figure-of-eight ribbon. Modern cardioid microphones work by allowing some sound to reach the back of the diaphragm.

Being partly figure-of-eight, cardioid microphones show a degree of bass tip-up, but this is not usually as much as in pure figure-of-eights. They are particularly useful in their ability to reject sound from behind them. It's probably true to say that there are far more cardioid microphones in use, world-wide, than any other type. For example, in television studios they can reduce greatly unwanted noise from, say, cameras; in multi-microphone music balances they can be placed close to the individual instruments and with many other instruments on the side of the microphone, where there is some reduction in pickup, and others at the rear of the microphone it becomes possible to achieve good separation of each instrument from the others.

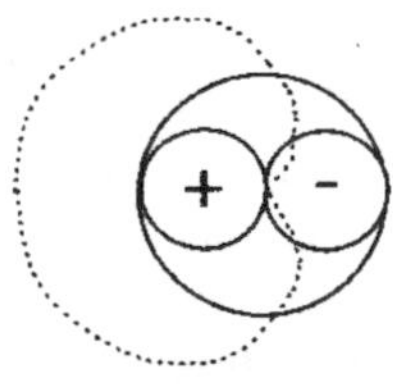

The cardioid (dotted) pattern is the addition of the other two. The front and rear lobes of the figure-of-eight have opposite polarities, so that the rear negative lobe has to be *subtracted* from the circle of the omnidirectional pattern.

We should point out that it is almost impossible for designers to produce microphones with the perfect cardioid shape at all frequencies. The *back-to-front ratio*, the difference in decibels between the front and rear axes, is rarely better than about 25–30 dB and may be as little as 5–10 dB at low frequencies.

4. *Hypercardioid.* This is a sort of half-way stage between cardioid and figure-of-eight. The 'dead' sides at 45° from the rear axis give yet another tool for the rejection of unwanted sounds.

And finally, there are:

5. *Highly directional or 'gun' polar diagrams*. Microphones with this characteristic take the form of fairly long, typically about half a metre, tubes with slots or perforations along the tube. Except at low frequencies, when they are more like cardioids, they have a fairly narrow *angle of acceptance*, which is about 60° at around 1 kHz, reducing to perhaps 30° at 7 or 8 kHz.

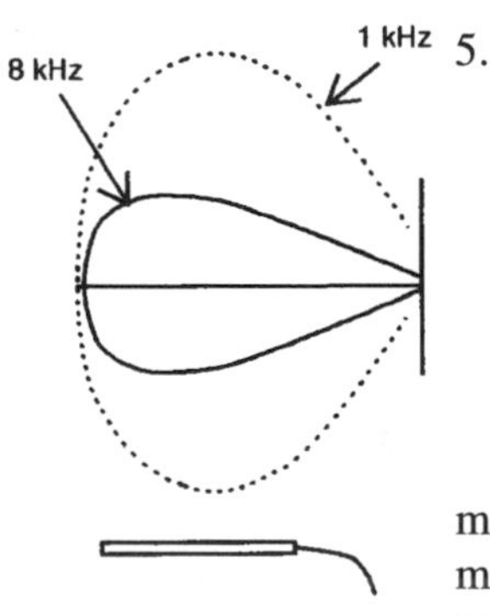

'Gun' microphone

We should mention the existence of *variable directivity* microphones. These are rather complex electrostatic microphones which work by having two cardioid capsules inside one case which face in opposite directions. By electrically *adding* the outputs of the two units, an omnidirectional response results, by *subtracting* there is a figure-of-eight pattern, and by 'switching off' one capsule a cardioid pattern is left. The reader might care to sketch two equal sized cardioids, placed so that their axes intersect at the same point, and prove to his or her satisfaction that what we've said is at least plausible.

9.4 Microphone applications

There are so many ways that microphones can be used that there's a serious risk of being misleading in giving a list of applications. Provided, though, that the reader accepts that what follows is no more than an outline, it is worth setting out broad guidelines.

Omnidirectional microphones

1. These are less affected by wind than other microphones and are therefore very suitable for out-of-doors interviews. Also they do not have to be pointed first to the interviewer and then the interviewee. It should normally be possible to hold the microphone still and between the two people.
2. Very small personal microphones of the kind clipped to the clothing are often omnidirectional.
3. Hand-held vocal microphones are usually cardioids but good results can be obtained with omnidirectional microphones.

Cardioid microphones

These are the most commonly used microphones and it's easier to list the things they are less well suited to than give their applications!

1. They tend to be prone to the effects of wind because there are slots behind the diaphragm to allow some sound in. Turbulence around these slots can be troublesome. However a good windshield can greatly reduce the problem.
2. They show a degree of bass tip-up, some cardioids more than others. This can be a drawback for close speech but more often it is turned to good account. A well-known hand microphone for vocalists, used close to the mouth, uses the bass tip-up effect to give a satisfactory quality, but used at a distance the bass output tends to be lacking.

Figure-of-eight microphones

These can be useful because of their ability to reject sounds from the sides, but they have their drawbacks. As mentioned earlier they are very prone to 'rumble' – a rumbling kind of noise is heard when they are moved – and they are also very susceptible to wind noise. However one well-known microphone, the so-called lip ribbon microphone, often used by sports commentators, has the ribbon very close (about 5 cm) from the mouth of the speaker. It is of course well-shielded from breath effects. Normally there would be excessive bass tip-up but a built-in electrical circuit corrects for this to give a reasonably flat response. Distant sounds, though, don't have the bass rise but they are still affected by the filter. The overall effect is that distant sounds, such as a football crowd, are greatly reduced.

Hypercardioid microphones

As might be expected these have the same drawbacks as figure-of-eights but to a lesser degree. The two dead sides, about 45° off the rear axis, can make this type of pattern useful in various applications where the unwanted noise is not directly behind the microphone.

'Gun' microphones

1. Television news gathering, where the gun microphone can be a reasonable distance from the action and not be seen by the camera.
2. Location drama, for the same reasons.
3. For picking up contributions by an audience in discussion-type programmes.
4. Gun microphones do not work particularly well in smallish rooms.

9.5 Microphone faults

Like any other item of technical equipment microphones can become faulty.

RULE ONE:
NEVER TRY TO REPAIR A MICROPHONE!

The reason is that they are complicated and delicate things and it's all too easy, unless you really know what you're doing, to get things out of alignment, overtighten a screw or in some other way to make things worse. Some of the fault symptoms, with remedial action where appropriate are given in the table on the next page.

Having decided that the microphone is faulty, and that the trouble is not just that of incorrect plugging, your supervisor will need to know the details, and it may be necessary for him or her to have it all written down:

1. What type of microphone (possibly its serial number as well)?
2. Where and when?
3. Describe the fault as accurately as possible. It isn't enough to say 'it sounds funny'!
4. State what checks you did.
5. Where is the microphone now?
6. Attach an identifying label to the microphone.

Symptom	*Possible cause*	*Action to be taken*
No output	1. Microphone not connected or plugged into wrong socket	Check plugging
	2. Not faded up on mixing desk	Check mixer
	3. If an electrostatic microphone power not present	Check powering
	4. Faulty cable	Replace cable
	5. Microphone may be faulty	Report to supervisor giving full details (see above)
Very low output	Incorrect setting on mixer	Check mixer settings
Output too high causing distortion	Incorrect setting on mixer	Check mixer settings
'Thin' sound with no bass	Faulty cable	Try a different cable
Excessive bass	Faulty microphone	Report to supervisor
Hisses and crackling noises in output	Damp in electrostatic microphone	Put microphone in a warm (but not hot) place. Check again in half an hour.

10

More about microphones

The reader might wonder why so much space is being given to microphones. The reason is simple. They are the only items that the user has a major degree of control over, in the sense that he or she can usually choose which to use for a particular purpose and can certainly decide on its positioning. All the other items involved in sound operations, loudspeakers, recording and replay machines and so on are for the most part already there and often in fixed positions. In a way they are more like items of furniture! Important, yes, but not coming under the direct and immediate control of the user.

10.1 Phantom power

In the last chapter we said that electrostatic microphones needed power, mainly these days for the pre-amplifier, although non-electret microphones need a fairly high voltage, usually in the region 50 to 100 V, for the capsule. Before transistors became thoroughly reliable a small valve amplifier was fitted inside the Microphone casing, and valves need a low voltage (about 6 V) for heating the cathode and a rather higher voltage, typically 50 V or more for the anode. All this meant that there had to be a fairly bulky power unit somewhere on the studio floor and a multi-core cable carrying these various supplies to the microphone and also to carry the audio signal back to the mixer. It was a messy, troublesome and expensive business!

With transistor amplifiers which need only one quite low voltage to operate them a more-or-less universal system known as *phantom power* has been adopted. It is illustrated in Figure 10. 1.

The important thing is that the two wires carrying the audio signal both carry the positive voltage (48 V). That means there is no voltage difference between them so that any non-electrostatic

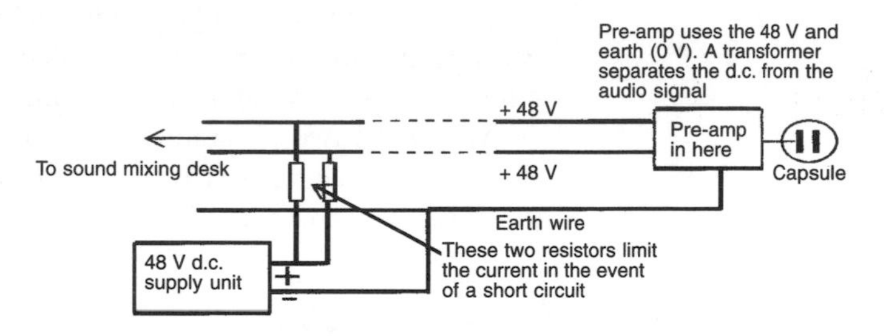

Figure 10.1 Phantom power

static microphone, a moving coil microphone for example, is quite unaware of this voltage. An electrostatic microphone, on the other hand, can extract what power it needs. Thus a standard three-core cable is all that's needed for any microphone. The power supply unit is usually incorporated in the mixing desk so that the phantom power is automatically available when a microphone is connected to it. Some desks have a switch – usually a small push button – allowing the phantom power to be switched off, so if an electrostatic microphone appears not to work the first thing is to check that the phantom power is switched on for that microphone.

A word of caution: plugging up any microphone when the phantom power is on can cause loud bangs in any loudspeaker connected to the mixer *unless the microphone has first been faded down*, by which we mean that its output has been reduced to nothing in the mixer. It's important to do this as a matter of routine.

There is one other common method of powering microphones. The one we have just described is known variously as *48 volt phantom*, or *standard phantom power*. This second type is usually called *A-B powering*. It's usually used for just one microphone at a time and we illustrate it in Figure 10.2.

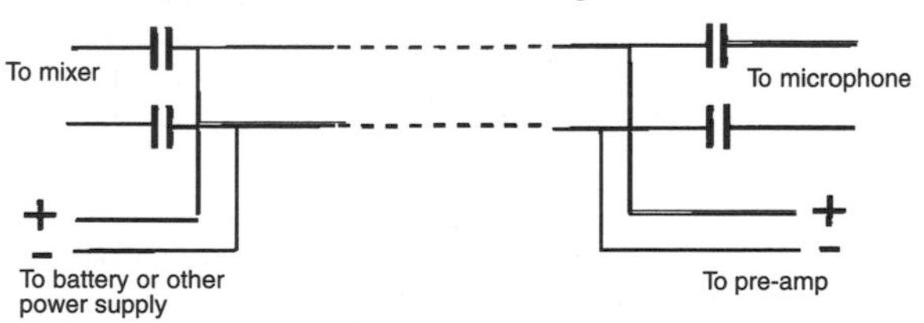

Figure 10.2 A-B powering

The voltage is generally low (9–12 V), and one advantage claimed for it is that it can still work if there is no earth connection. (But this is likely to be unfortunate because it's the earth connection which helps to reduce the effects of induced hum and other unwanted electrical interference.)

Also if the two programme wires become changed over in their pin connections, causing a *phase reversal*, then A-B powering won't work.

10.2 Phase in microphones

We met the idea of 'phase' when we were dealing with a.c., and we have mentioned it in the preceding section. This may be an appropriate point to cover the subject again, but in a slightly different context. (Phase will crop up again in Chapter 12 on stereo, although in a slightly different sense.)

At the moment let us imagine a pair of exactly similar microphones placed in front of a speaker – a common enough situation. If the speaker is an eminent person then it's only sensible to have a standby microphone, or maybe one might be used as a feed for one network, the second microphone for a different network. It will be almost the same distance from each microphone to the speaker's mouth, so their outputs will be virtually the same. Now suppose the connections to one microphone were reversed compared to the other. Then if *both* microphones are faded up together, and they are in the same circuit, the outputs will, in theory, cancel, giving no output, as illustrated in Figure 10.3. It's therefore very important to check before the programme or the recording that the *phasing* of the microphones is the same.

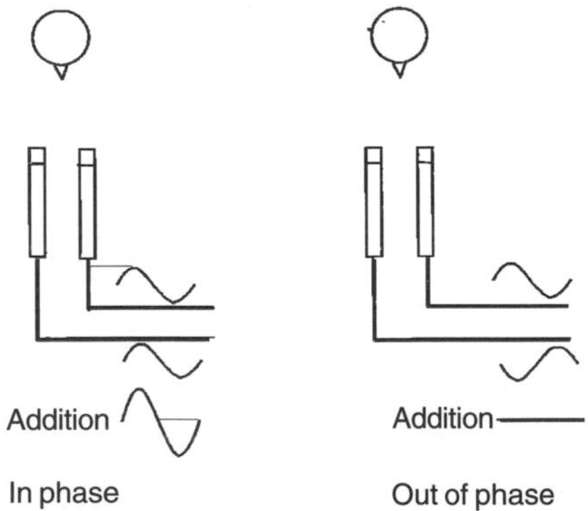

Figure 10.3 Microphones in and out of phase

The procedure is quite simple. We need two people, one to speak into the microphones, ensuring that he or she is equidistant from both, and the other to operate the sound mixing desk.

1. Fade up microphone 1 to give a normal reading on the meters (more about those later). Note that reading and fade out microphone 1.
2. Fade up microphone 2 until it gives the same reading.
3. Now fade up both microphones. If they are correctly phased the meter readings will increase by 6 dB (the reader should understand why it's 6 dB).

If one microphone is out of phase compared with the other then the output will drop and theoretically there should be complete cancellation of the sound. In practice this never happens, but the sound will certainly be at a low level and 'tinny' in quality.

This test doesn't tell you which microphone (or cable) is wrongly phased, but at least it gives an opportunity to prevent a possibly disastrous situation.

10.3 Microphone cables, plugs and sockets

We mentioned the cabling to microphones in Section 10.1 on phantom power. Now is the time to say a little more. The professional world has for many years standardized on one type of microphone plug and socket, known as 'XLR', a typical plug being shown in Figure 10.4.

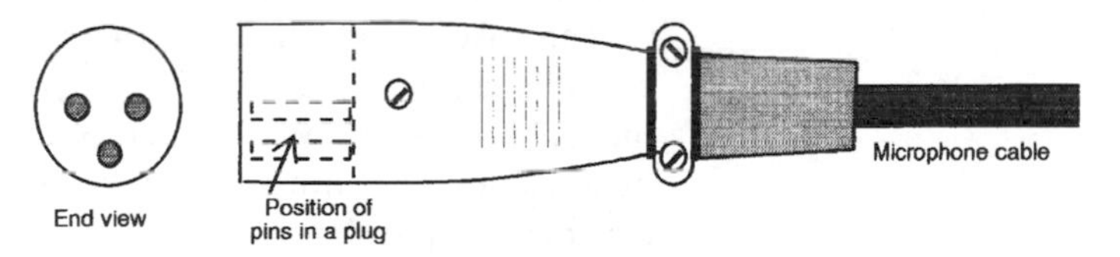

Figure 10.4 An XLR plug on the end of a cable

XLR plugs are found on microphones so that it is a *socket* which is on the end of the cable connected to it. It is useful to remember which end of the cable goes where – it's all too easy to lay out a long piece of cable tidily and then find that it's got to be turned round! Think of it this way: most studios have assemblies of various types of connection points – for microphones, mains, cameras, and so on – at places on the walls. If microphone points had plugs with pins these could perhaps get damaged, so invariably sockets are used. Therefore the microphones themselves must have plugs with pins. (This might not be the original reason for the convention but it's an easy one to remember.)

Then the cable itself. The very small voltages coming from microphones are prone to interference from other electrical devices, particularly in a television studio with cameras and, most important of all, lighting circuits. The latter often carry large currents and modern lighting control equipment works by, as it were, 'chopping out' part of the mains waveform. This process creates large numbers of harmonics which can radiate out from the wiring. Consequently the conductors carrying the audio signal from the microphone need to be screened, and the simplest way of screening two wires is to make the earth (or return wire) in the form of a flexible screen surrounding the other two.

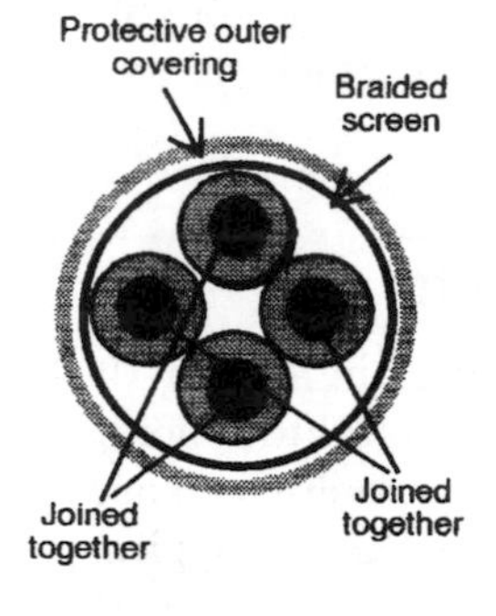

Star-quad cable

Even this is often not enough in television studios so a further form of cable is used known as *star quad*. In this there are four conductors inside the screening and opposite wires are connected together. Not only that but the four conductors are twisted round each other. Now in ordinary three-core cable inevitably one programme wire is slightly closer than the other to the source of interference. With star quad the arrangement is more symmetrical, or *balanced*, in electrical terms. This means that the induced voltages are virtually the same (but opposing) in both programme carrying conductors. They therefore cancel each other, making this type of cable much less prone to the effects of interference.

This takes us to the subject of balanced wiring in general and this applies to all kinds of wiring, not necessarily just microphone cables. We have seen a good example with star-quad cable. The aim is to make both sides of any circuit electrically equivalent to each other. This may involve in some instances making the two conductors take mirror-image routes.

10.4 Variable polar diagram microphones

In the last chapter we said that some microphones could have their polar diagram changed. Such microphones are, not surprisingly, expensive, but they can be extremely useful. We will explain how this is done, but the more mathematically-minded reader might find the section in the box interesting.

We will let r represent the sensitivity of the microphone at the angle ϕ.

1. Omnidirectional microphones. Here, at least at the lower frequencies, r is independent of angle, so we could write $r = 1$.
2. Figure-of-eight. Since $\cos 0° = 1$, and $\cos 90° = 0$ it should not be too surprising to find that $r = \cos \phi$.
3. We said, and showed diagrammatically, that a cardioid pattern is a combination of omni and figure-of-eight patterns, so for a cardioid microphone $r = 1 + \cos \phi$.

The reader with a scientific calculator handy can easily verify that these equations are at least plausible!

Now the variable polar diagram microphone is made up with two cardioid units, almost invariably electrostatic, although in the past there have been attempts to use moving coil units. The two cardioids are back to back, and in fact they are not separate capsules but an ingenious use of one back plate, perforated to allow some sound waves to get through, and with two diaphragms, one on each side. It's important that the electrical voltage can be varied, so these can't be of the electret type – there has to be a potential of, typically, 50 V between the diaphragm and the back plate. Figure 10.5 shows the basic circuit.

If the rear diaphragm is at 50 V, as it is in the diagram, then the rear cardioid is inoperative because its diaphragm and the plate are at the same voltage. Consequently there is just the front-facing cardioid.

If the switch is set to the lower contact, at 0 V, then in effect we have both cardioids working in the same phase so they are added together and this gives an omnidirectional response. This could be taken on trust but a more mathematical explanation is given in the box overleaf.

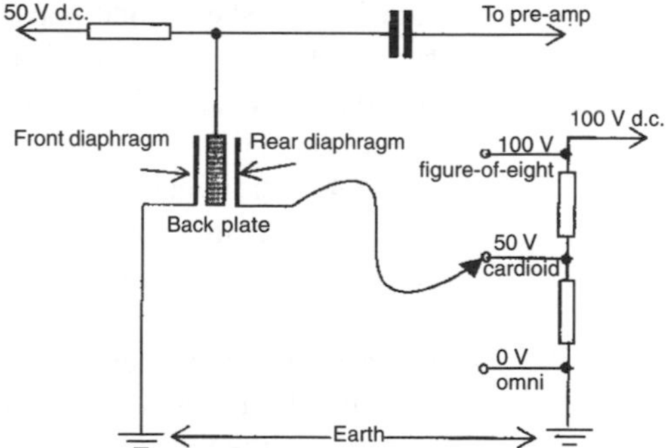

Figure 10.5 Circuit of a variable polar diagram microphone

Now, with the connection to the 100 V point, the two cardioids are in opposition to each other so we subtract one from the other, and this gives a figure-of-eight.

And, with more resistors in the chain it's possible to have other diagrams, including hypercardioid. The one that can't be produced in this way is a 'gun' response.

An interesting exercise that the reader might care to do is to show by simple sketches that adding and subtracting cardioids results in the polar diagrams we have deduced.

Adding two cardioids together mathematically we have:

$$1 + \cos\ \phi + (1 - \cos\ \phi) = 2,$$

which is a constant, implying an omnidirectional response. (Notice that we had to put a minus sign in for the rear cardioid to show that it was facing in the opposite direction.)

Subtracting cardioids we have:

$$1 + \cos\ \phi - (1 - \cos\ \phi) = 2 \cos\ \phi$$

which is a figure-of-eight response. Admittedly there is a 2 there, but that doesn't affect the shape of the polar diagram.

10.5 Other sound pick-up devices

1. *Pressure zone microphones*. (PZMs). These are not really anything radically different from the microphones we looked at earlier. A PZM is simply a different way of using a microphone. In essence they are a conventional microphone, almost always an omnidirectional one, mounted in a suitable block of, say, wood, or fixed close to a flat metal plate. The idea is that when sound waves are reflected from a hard surface there is an effect called *pressure doubling*. Thus, a microphone placed close enough to such a surface should experience a 6 dB increase in level.

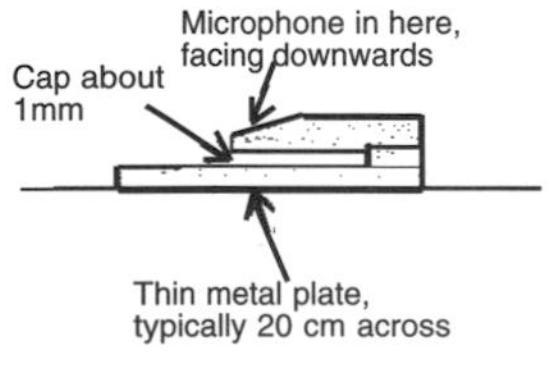

Pressure zone microphone

PZM microphones in one form or another have been found useful in a number of applications. One is as a 'footlights' microphone for stage performances, when one or more are placed on a stage near the front, where they are inconspicuous; another is as a pick-up device for round-the-table discussions. It may be reasonably fair to say that PZMs have their devotees amongst sound balancers. Those who like them use them wherever possible. Others prefer more standard methods of sound pickup.

2. *Direct injection (DI) boxes.* These ate not microphones at all but devices for connecting to electronic musical instruments. A typical use is with electric guitars where, for example in rock music, every instrument has its own amplifier and loudspeaker. It may be very difficult to place a microphone in such a position that it picks up little of the other instruments. DI boxes are connected between the instrument and its amplifier and the DI box has, typically, an XLR plug like a microphone so that a microphone cable can go from it to the mixer.

   Some DI boxes need a power source, others are *passive*, meaning that they don't need power. The important thing is that they provide electrical isolation between the instrument and the mixer, especially in a safety sense.

10.6 Microphone stands and cabling

The variety of these is very considerable. Important characteristics are that they should be stable and not move when locked in position and that any vibration in the floor or table is not transmitted to the microphone. They may also need to be capable of being folded up for tidy storage.

Microphone stand with boom arm

Among the types are *table stands* which are self-explanatory, floor stands, which are usually adjustable in height from perhaps 1.5 m to more than 2 m, and *floor stands with a boom arm*. These are particularly versatile as the microphone can be placed over an instrument or a table, lowered to a short distance above the floor and raised to perhaps 2.5 m.

On the subject of cabling, we can say that the manner in which a microphone, or any other piece of equipment on a studio floor has its cable laid out, shows whether a truly professional person has been at work! To begin with safety is vital. Cables should never be laid in a way which could cause anyone to trip over them – and this is especially important when the public are going to be in the area.

Secondly the cabling should look neat. This applies very much to television, for obvious reasons, but untidy cables can create a bad impression in the minds of performers and public even when there are no cameras present. Thirdly, a neat layout makes it easier to track down and remove a faulty cable. And finally such a layout makes the *de-rig* quicker and easier at the end of the show.

If things like microphone cables have to cross say a doorway, they should if possible go over the doorway If that isn't possible then they should be taken under proper cable guards, or if those aren't available then they should be taped to the floor. In an emergency a piece of carpet or a door mat should go over them. Microphones in a stand should have the cable neatly clipped or taped to the stand right down to the floor. And table microphone cables should be taped to a leg of the table, again down to floor level. (It's sensible not to do all this fixing and taping until you're sure that everything is working – so that cables don't have to be replaced and microphone stands have been set to their final positions.)

Always think **safety** when rigging equipment

Of course, all this is perhaps a little easier said than done if it's television, but the good sound operator is constantly aware of the need for neatness, not only for its own sake but also in the interests of safety both for people and equipment.

10.7 Microphone placing

An impossible subject to deal with in print – or by almost any other means of communication! Still, we can set out a few guidelines, but it has to be said that no two experienced sound balancers are likely to agree *totally* with each other on this subject. There are individual tastes and no two programme situations are ever going to be exactly similar. Nevertheless. . .

First there is the matter of *sound perspective.* This means the apparent distance of a sound source as judged by the ear. Sometimes the sound perspective doesn't matter. At other times it does. For example, an actor in a play, if supposed to be at the far side of a room, ought to sound more distant than other actors near at hand.

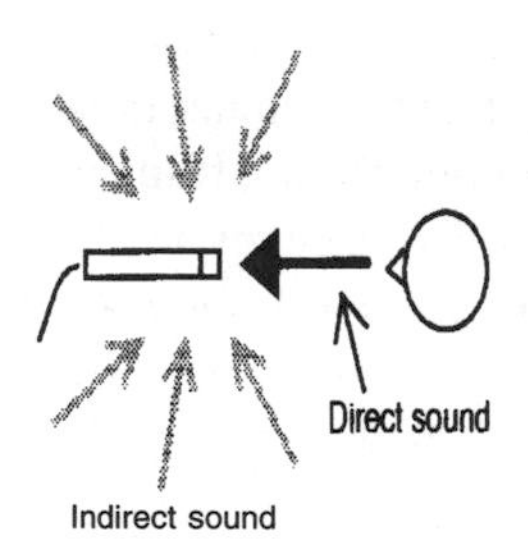

Again, with classical music, it can be important to make the instruments at the back of the orchestra sound further away than those at the front. But there are no hard rules about when perspective is important and when it isn't.

What determines sound perspective? Various things, such as loudness and quality can sometimes have a bearing but the most important fact by far is the ratio of the *direct sound* to the *indirect sound.* The direct sound is that which goes straight front the source to the microphone. The indirect sound reaches the microphone(s) having undergone reflections from walls, floor, ceiling and so on. It is the *reverberant* sound reaching the

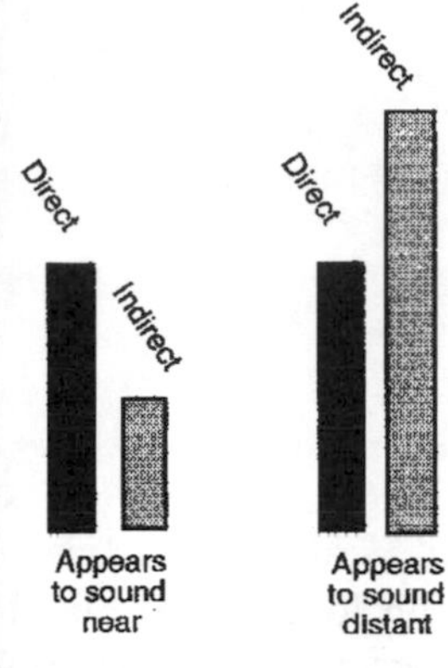

microphone. If there is little indirect sound then the source seems near, if there is a lot of indirect sound, compared with the direct, then the source appears distant.

As we've said, sometimes perspective matters very little, but it can be extremely important on other occasions. If the reader has never noticed sound perspectives it may be that the reason is that they've been correctly handled all the time! It's quite instructive to listen hard to, for instance, a radio play trying not to concentrate on the plot but paying attention only to the apparent distances of the actors. It's also interesting to listen intently to television but without looking at the screen. (It may help to hang a sheet of newspaper over it.)

With the human voice the best place for a microphone is likely to be directly in front of the mouth at a distance which might be anything from a few centimetres to a metre. This is fine for radio, but generally for television, unless it's a vocalist holding the microphone very close, this is visually not acceptable. The distance and the acoustics may vary but in general speech pick-up is acceptable if the microphone is within about 45° of the axis of the mouth.

And finally, musical instruments. Sometimes, and this is mostly in the realm of classical music, it is important to capture the true sound of the instrument. There is one good rule here (but like all rules it may have to be broken occasionally): a microphone should look at the instrument from the direction of an audience. Trumpets are usually played with the bell out to the front; put the microphone in line with the bell. Clarinets are usually pointed downwards to some extent. The audience isn't looking into the bell of a clarinet so a microphone shouldn't. French horns have their bells pointing backwards towards a reflecting surface: the microphone should be pointed towards the reflecting surface, and so on.

But remember that what we've said about placing microphones is no more than a general guide which holds reasonably true for much of the time but is by no means infallible!

11

Loudspeakers

In this chapter we shall concentrate on loudspeakers used for *monitoring* – that is for assessing accurately by ear the audio signal being recorded or transmitted. Some of what is said may apply to high grade domestic loudspeakers, but not necessarily so.

11.1 The parts of a loudspeaker

It's helpful to see loudspeakers as being like microphones except that they emit sound instead of receive it. They have similarities in that, like microphones, there are three principal parts and these correspond quite closely to the parts of a microphone. (In fact in many intercom systems the small loudspeakers are also the microphones.)

First, corresponding to the diaphragm, is a *radiating surface.* It may not normally be called that, but that is what it is. Normally it's a cone whose vibrations produce the sound waves. Secondly there is the *transducer*, which in this instance converts electrical energy into mechanical movement, and thirdly there is the *enclosure*, or cabinet. Now we saw that with microphones the case affected the polar diagram of the microphone. That could happen with loudspeakers, but here the situation is rather different because the nature of the box has a profound effect on the frequency response, and in particular the bass.

As with microphones we'll take these three parts in turn, except that it's not very helpful to lump the transducer and the radiating surface together. We could do that with microphones because often the two were integral – the diaphragm being part of the transducer.

11.2 The radiating surface

In almost all monitoring loudspeakers this is a cone-shaped structure, so in future we'll simply call it the 'cone'! The job of the cone, driven by the transducer, is to create sound waves, which sounds simple. In fact it's far from simple because it isn't easy to design a cone which is large enough to create the loud sound levels which are often required and to do this smoothly from 30 Hz or perhaps less, up to at least 16 kHz, and at the same time avoid introducing distortions or *colourations*, the latter being frequencies which were not present in the original audio signal.

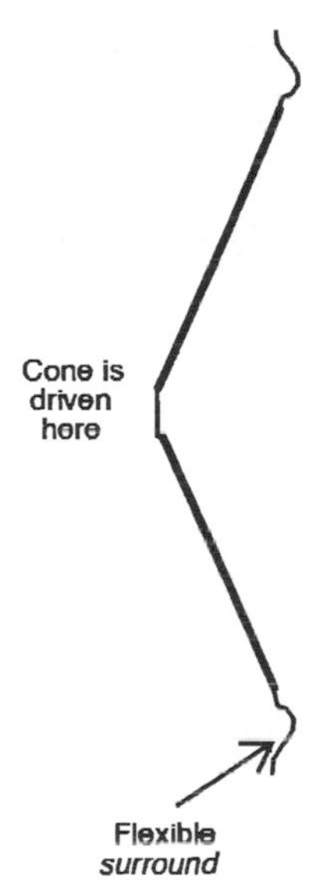

Low-cost cones are often made from a kind of fibrous compressed paper, but for professional loudspeakers various plastics are favoured, among them one called *polypropylene*, which can be shaped by drawing the heated material down over a mould by a vacuum. Although it may not seem so at first sight the cone and its design is really quite complicated. Apart from all else it must not have its own modes of vibration. It's all too easy for a cone, driven near its centre, to move correctly to and fro at low frequencies but there is a tendency at high frequencies for conical things to behave like a bell, for example, and vibrate in curious ways. And all the time the cone has to be capable of moving enough to produce quite a lot of noise. It's not too difficult to make a cone of moderate size, up to about 15 or 20 cm diameter, work reasonably well over much of the audio range, but to produce high sound levels such a cone needs to make large amplitude movements, and that may not be easy to achieve with standard types of transducer. Large diameter cones, 30 cm or more, tend to go into peculiar modes of vibration at the higher frequencies, so the answer, which we shall go into later in this chapter, is to have *multiple unit* loudspeakers, one large cone assembly for low frequencies, and one or maybe two more smaller ones for the higher frequencies.

11.3 Transducers

As with microphones there are many possible types of transducer but only one is widely used in monitoring loudspeakers – the *moving coil unit*. In principle this is very similar to the moving coil microphone, but bigger and more robust. Figure 11.1 on the next page is a simplified diagram.

One important aspect in the design is to ensure that the coil is always in a linear part of the magnetic field. That is, the flux density must be constant over the whole length of the coil, no matter how far it moves in or out. There are two, at first sight contradictory, ways of doing this. One is to have the coil very short, so that it is always well within the length of the gap; the other is to have a very long coil so that even if the flux density is not constant the coil extends well beyond the non-linear region. In other words the coil, is always in the *same* field, even if that field isn't constant within itself.

11.4 The enclosure

We've already implied that the casing of a loudspeaker was very important, especially in the matter of the low frequency response. We'll see why.

First, imagine a *drive unit*, that is, the transducer and cone considered as one piece, fed with a suitable signal, but not in any sort of case.

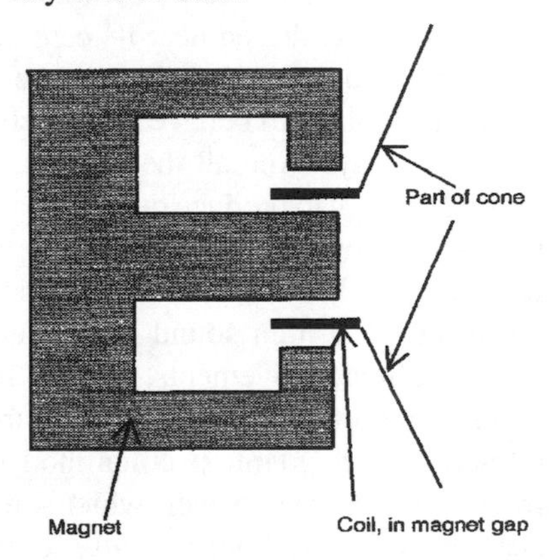

Figure 11.1 Moving coil loudspeaker unit

We have to remember the rules about diffraction, and think of the front of the cone as being an aperture. The high frequencies will radiate out from the front and also from the rear of the cone, but the low frequencies will bend round, from the front to the back and from the back to the front. So, if we imagine a point some little distance in front of the unit, there will be a reasonable amount of high frequency but there will be very little low frequency energy because we shall be receiving sound waves from both the front and the back of the cone, and *they will be*

out of phase! This is because when the cone moves forwards it creates a compression at the front but a rarefaction at the rear. The overall result is a thin and 'toppy' sound. Readers with an old loudspeaker unit removed from its cabinet (an old radio set for example) might be able to verify this for themselves. It should be clear that it is the sound which comes out from the rear of the cone which causes the nuisance and it is the main function of enclosure designers to get rid, somehow, of this rear radiation. A first, and partially satisfactory answer is to mount the drive unit in a *baffle*, a large board with a hole in it.

This idea has been used and has been found to be fairly satisfactory for speech-only loudspeakers. The difficulty is that to prevent the rear radiation from diffracting to the front the baffle must be a few wavelengths across. At 30 Hz λ is around 10 m, so a baffle 30 to 50 m across would be needed! The speech-only baffles mentioned above were about 1 m square and Figure 11.2 shows the sort of frequency response of such a device.

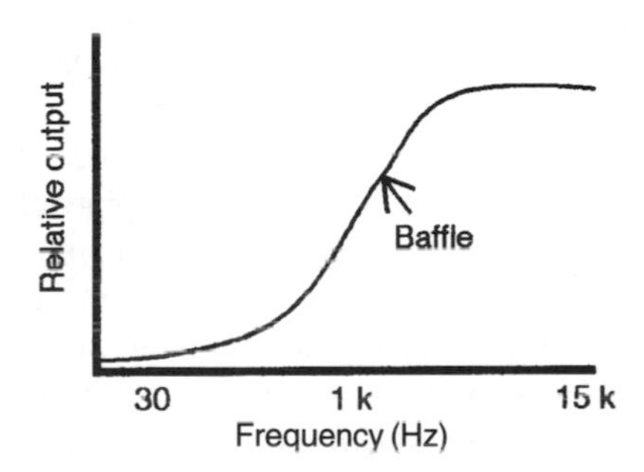

Figure 11.2 Response of a 1 m square baffle loudspeaker

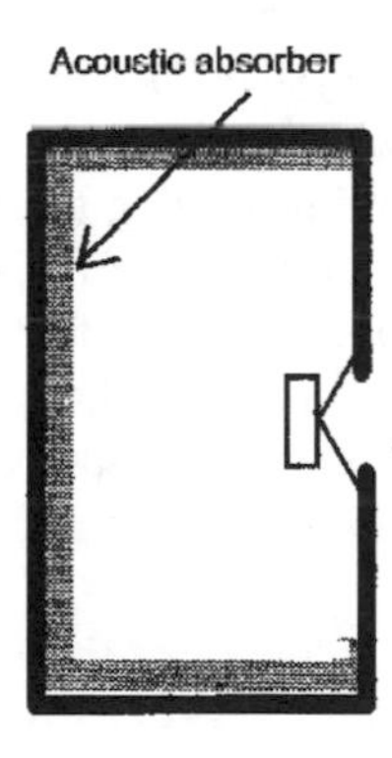

A better idea is to put the drive unit into a sealed box. This certainly stops the rear radiation and some acoustic absorbent material reduces the possibility of standing waves inside the box. Unfortunately the sealed enclosure turns out not to be the perfect answer because, unless the volume of the box is very large, several cubic metres perhaps, the cone in moving backwards is trying to compress the air, and this restricts the motion at low frequencies. A lower limit of around 60 Hz results. Despite this, sealed enclosure loudspeakers are used in the professional world where there are difficulties like a shortage of space so that only small units can be used. Figure 11.3 shows the kind of response that a sealed enclosure loudspeaker can have, with the baffle-only response shown by a dotted line for comparison.

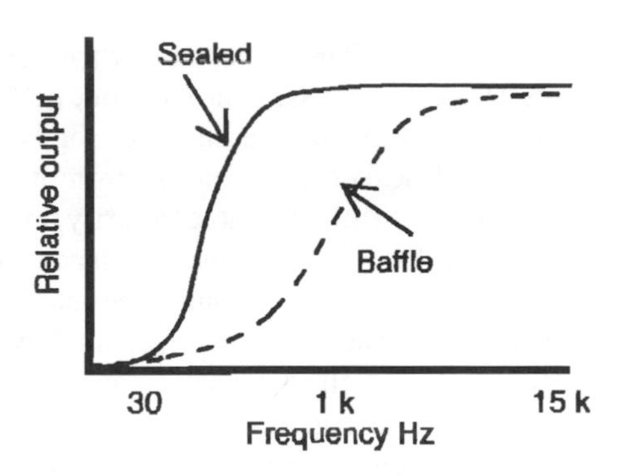

Figure 11.3 Response of a sealed enclosure loudspeaker

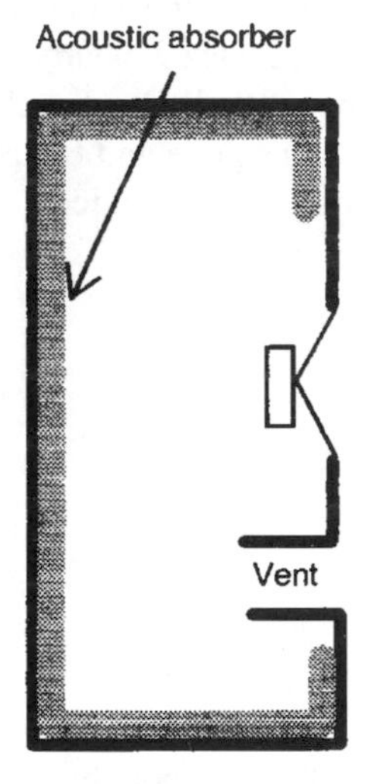

A really good response at the bass end involves a radically different approach to the problem. The *vented enclosure* loudspeaker has, to begin with, a fairly large enclosure but it is not just a box – it is a resonator! (Vented enclosure loudspeakers are sometimes called *bass reflex* loudspeakers.) The reader is invited to blow across the neck of an (empty!) bottle – whisky, gin or beer – and notice that the note produced is a low frequency one. Many gin and whisky bottles produce a note of frequency about 110 Hz. This is much lower than the frequency which results from blowing across a pipe, closed at one end and of the same length (30 cm or so). The explanation, perhaps a surprising one, is that the mass of air in the neck, small though it is, is 'bouncing' against the volume of air in the bottle. Resonators of this sort find a number of applications and they are known as *Helmholtz resonators*, after the German scientist who apparently first made use of them.

Now the fact that the neck or vent in such an enclosure goes inwards has no effect on the working of the system. The designers ensure that the resonant frequency is low, usually around 30 Hz and it becomes possible not only to avoid restricting the cone movements at low frequencies but even to boost the response where the drive unit might be becoming less efficient. Interestingly, the air movements in the vent are in phase with the cone movements! Figure 11.4 shows a typical response for a vented enclosure compared with a sealed enclosure.

11. 5 Multiple unit loudspeakers

We said earlier that it was difficult for a single cone of more than about 30 cm diameter to handle a wide range of frequencies

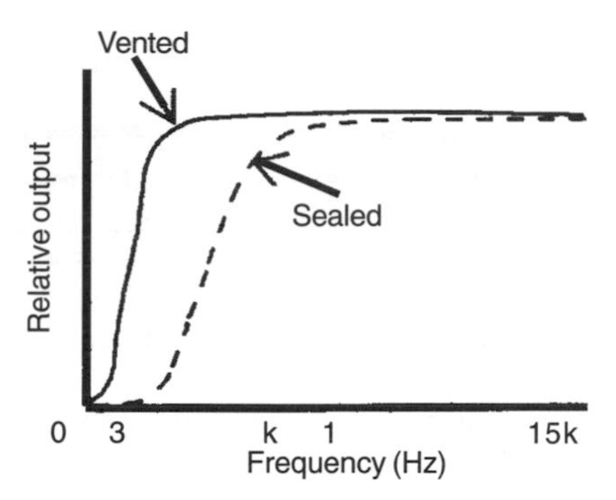

Figure 11.4 Response of a vented enclosure loudspeaker

and at the same time produce reasonably high sound levels. (High sound levels in a monitoring loudspeaker mean something like 115 to 120 dBA at 1 m in front of the loudspeaker.) The solution here is to have more than one drive unit:

1. A small one – the *tweeter* – typically handling frequencies above 2 kHz.
2. A *mid-range unit*, covering from several hundred Hertz up to the start of the tweeter range.
3. A large *bass unit*, dealing with the frequencies below those of the mid-range unit.

The system is illustrated in Figure 11.5, which shows that the three filters form what is known as the *crossover unit*. The *high pass filter* allows only the higher frequencies go though it.

The *band pass filter* lets a band of frequencies, appropriate to the mid-range unit, pass, while the *low pass filter* deals with the part of the signal intended for the bass unit. This arrangement can work extremely well but there are problems. To begin with, the design of crossover units is complicated if there is to be a smooth transition from one band to another, as in Figure 11.6.

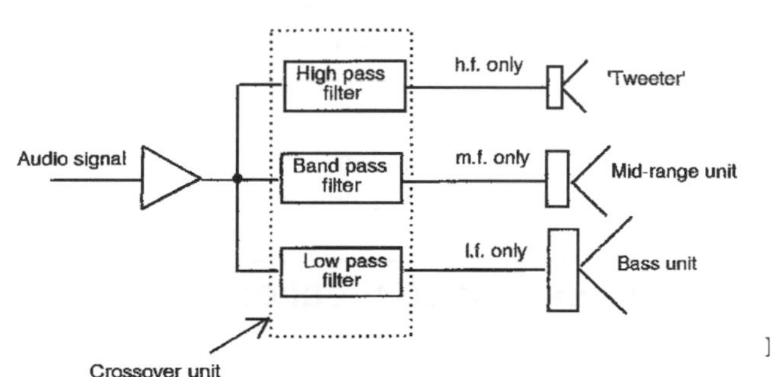

Figure 11.5 Multiple unit loudspeaker

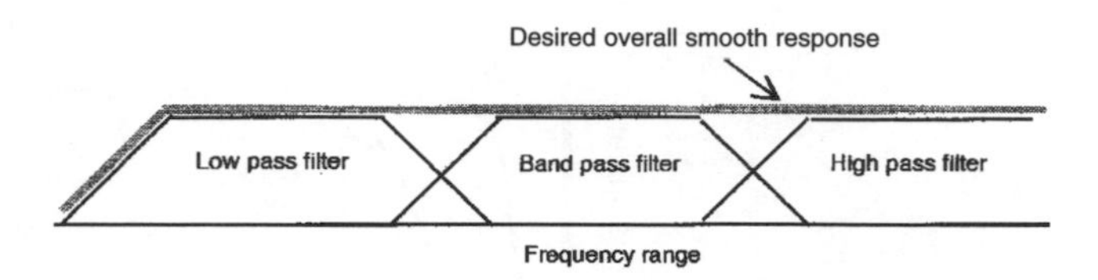

Figure 11.6 Frequency ranges in a crossover unit

Secondly, the circuits in the crossover may have to handle relatively high currents and/or voltages, since they come after the power amplifier. This makes design more difficult if the characteristics of the circuits are to be stable – the values of, in particular, resistors are apt to change as the currents in them may be high enough for an increase in temperature to occur.

An interesting and effective answer has been to develop what may be termed the *bi-amp* system, illustrated in Figure 11.7.

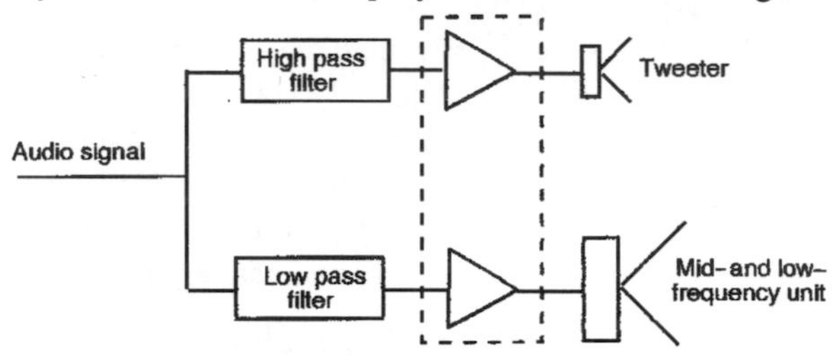

Figure 11.7 Bi-amp system

It is possible with careful design to cover the full audio range with only two drive units. Each unit can then have its own power amplifier and the crossover units can operate at ordinary audio signal levels. This simplifies the crossover circuitry and makes it more stable. Further, the two amplifiers can be the two halves of a stereo amplifier, which leads to compactness and lower costs. The low-level crossover system has a number of advantages: the total cost is no higher although there is a stereo amplifier because cheaper components can be used for the crossover units, there are only two of these, so design is simplified and the frequency response does not vary with the amount of power going to the drive units.

11.6 Directional loudspeakers

There is no such thing as a really directional loudspeaker! However there are units called *line-source* or *column*

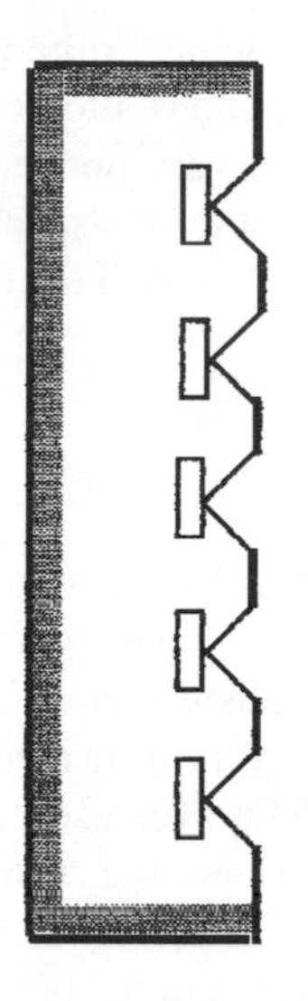

Line source (column) loudspeaker

loudspeakers which have sufficient directional properties to be useful. They get their names because they are a line of sound sources, or they form a column of drive units.

Figure 11.8 shows the principle. At a point X out on the axis of the column it should be obvious that the sound waves from the column will all be in phase and therefore reinforcing each other. AA^{I} and BB^{I} represent the paths of sound waves going out at an angle from the stack and there is clearly a path difference of *d*. This is going to mean a phase difference and therefore an element of cancellation in the sound waves at this angle, and if d happens to be a half wavelength then there will be total cancellation. Overall there will be a reduced sound level, *provided that the sound wavelengths are comparable with or less than the length of the loudspeaker column.* Of course, this directional effect is only going to occur in the same plane as the stack: if the column is vertical then the directional effect is also in the vertical plane. The little side diagrams show this. The loudspeaker array is more or less omnidirectional in the horizontal plane.

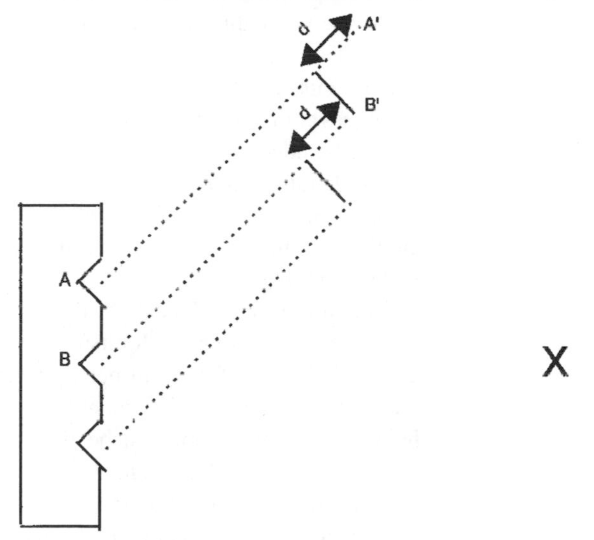

Figure 11.8 Principle of a directional loudspeaker

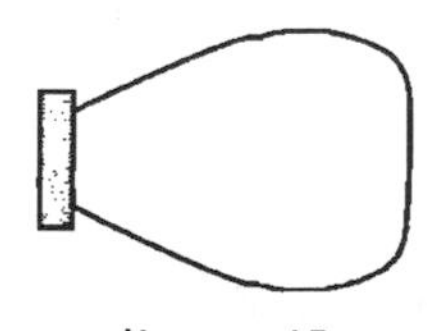

Line source LS polar diagram, side view

With many installations the audio signal fed to column loudspeakers has the bass reduced below about 160 Hz, so avoiding any output in the region where the unit loses its directional effect. This in turn means that line source loudspeakers are really only suitable for speech. (High quality units do exist but they tend to be large and expensive.)

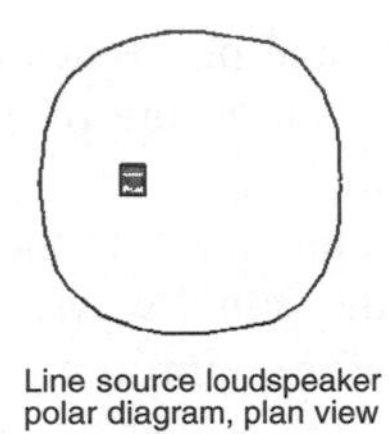
Line source loudspeaker polar diagram, plan view

Although the directional effect is somewhat limited line source loudspeakers can be very useful in sound reinforcement ('public address') work, where there can be very unfortunate effects if a microphone picks up enough of the loudspeaker's output to cause an oscillation, or howlround. We shall deal with this more in a later chapter.

11.7 Cables and loudspeaker positioning

The wiring used for loudspeakers really requires little comment. The important thing is that it must have a diameter which is adequate for the job. For example if a loudspeaker is being supplied with 50 W of power, which may be very little in a rock concert but might be typical of the audio power used for a monitoring loudspeaker in a studio control room, the current is likely to be of the order of 3 to 5 A. Too thin a cable will mean that heating occurs with a consequent reduction in the power to the loudspeaker, and this may be especially important if there is a long run of cable, although we shall later see that there are ways of reducing this effect by having a higher voltage and a smaller current. As a rough guide it could be said that a cable thickness equivalent to that used for mains cable of 13 A capacity should be adequate.

Incidentally, the reader is warned not to take too much notice of statements to the effect that the direction the loudspeaker cables go is important, and that there can be a change in quality if the cable is reversed!

The *siting* of loudspeakers is something that the user often has little control over. A loudspeaker should not be placed too close to corners of a room because there can be reflections of the sound from the walls causing interference effects. Also the height above the floor can be important. There have been a number of monitoring loudspeakers which have an optimum height above floor level of about half a metre, and special stands have been designed for them. Unfortunately that kind of height is often a very inconvenient one in control rooms as it is below the level of the sound mixing desk!

11.8 Headphones

Headphones can be thought of as miniature loudspeakers, although headphone designers would doubtless be very unhappy to have their creations dismissed as lightly as that! What we can only do here is to give a few words of caution about their use.

To begin with, serious monitoring with headphones is rarely as satisfactory as it is with loudspeakers. The reasons for this are perhaps partly psychological, although when we deal with stereo we shall see stronger reasons. It appears all too easy for a person wearing headphones to be isolated from the rest of the production team and a sound balance obtained with headphones is apt to be rather different from one obtained by loudspeaker monitoring. Sometimes, of course, headphones have to be used. Much location work, obviously, has to be done with them, but when there is a choice loudspeakers should be preferred.

12

Stereo

12.1 What do we mean by 'stereo'?

The word *stereophony*, invariably shortened to stereo, means 'solid' or 'three-dimensional hearing'. In fact that isn't a very accurate definition. 'Stereo', as we have come to understand the term, is really two-dimensional, although there can be an illusion of depth. Furthermore, it is universally accepted at the present time that stereo images are conveyed by two *channels*, that is, a system of two (or more) microphones, two sets of wiring to a recording device which itself has two tracks, and so on, to end up at two loudspeakers. There are arguments to the effect that more than two channels are desirable, but so far it has been proved that very satisfactory results can be obtained with no more than two. Also more than two adds enormously to the cost. So here we are going to deal with two-channel stereo. To see how stereo works we must first look at how we locate sounds in real life – what is known as *binaural* hearing.

12.2 Binaural hearing

A number of things are involved in our location of the direction of a sound. Common sense is one – if, in a sitting room, we hear a bird singing we shall expect the sound to come from the direction of an open window and not from a bookcase! What we see can also affect our perception of the sound's direction, and a further complication is that we can be conditioned to associate particular directions with certain sounds. There is, though, one basic effect which is a very powerful influence, and fortunately it is one that can be made use of in stereo reproduction. It is the *time-of-arrival difference* at the two ears. We shall abbreviate this to 'TOA difference', and Figure 12.1, on the next page, illustrates it.

If a sound is directly in front of the listener then the sound will enter each ear at the same instant and the TOA difference will be zero. However, for sound entering from the direction of X there will be a time difference corresponding to the extra distance *d.* Such TOA differences are small. In the extreme case when sound arrives from the side, at Y, the TOA difference for the average-sized head is less than a millisecond! And even more surprising is the fact that most people can detect a slight change in direction if the sound source moves from directly in front to 1° to the side. And then the TOA difference is less than 10 μs!

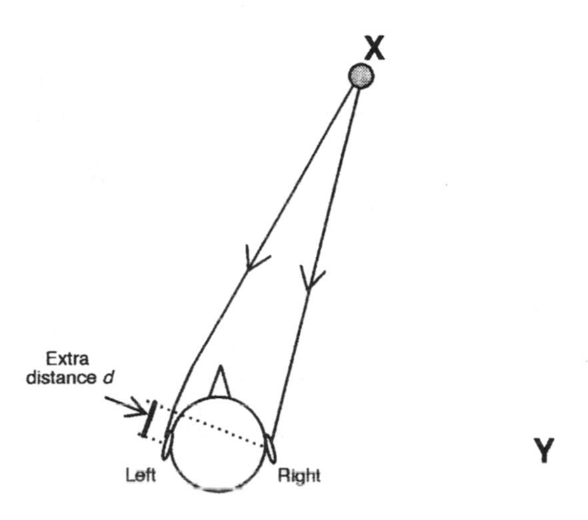

Figure 12.1 Time of arrival difference

For stereo to work successfully, then, there must be a way of simulating the TOA differences. In the early days of stereo, back in the 1950s and before, it was thought that the answer was to have two microphones spaced well apart in front of the performers. This did not work at all well, for reasons which we shall see. The first person to see the proper way of producing stereo was an Englishman named Alan Blumlein who took out patents on the subject as long ago as the 1930s. Blumlein said that what was needed was a pair of directional microphones, cardioids or figures-of-eight, as close together as possible and angled at 90° to each other. And that works well although, as we shall see, it's not the only way of producing stereo signals.

The possibly surprising thing is that if there are *amplitude* differences between the signals in the two channels these become, with a pair of loudspeakers, the equivalent of *time of arrival differences* at the ears. To understand this we first have to appreciate that if a listener is sitting in the optimum stereo listening position – on the centre line between the speakers and positioned so that the loudspeakers and the listener form an equilateral triangle – then each ear receives sound from both speakers (diffraction again!). Figure 12.2 helps to make this clear.

Let us now look at what happens at one of the two ears – the left one for instance. Suppose the amplitude of the sound from the left loudspeaker (LS) is rather greater than that from the right. As the upper drawing in Figure 12.3 shows, the larger sound wave from the left LS arrives at the left ear before the weaker one from the right LS. What actually enters the ear is the addition of these two sound waves, as in the lower diagram.

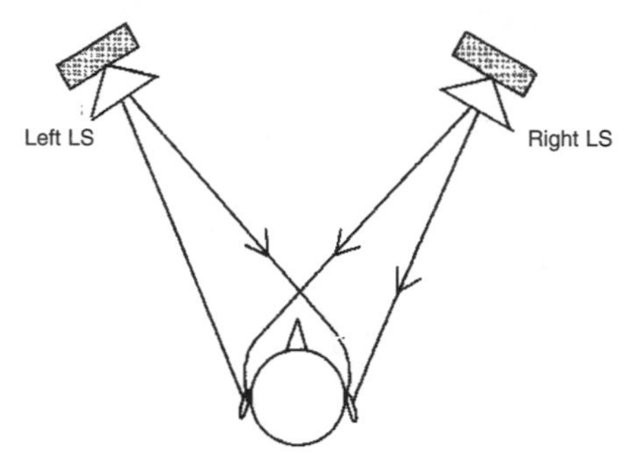

Figure 12.2 Sound from each loudspeaker enters both ears

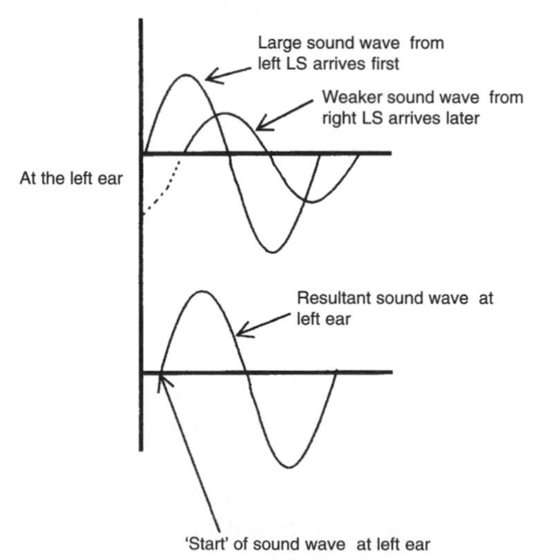

Figure 12.3 Creation of a TOA effect

It will be seen that the resultant wave has an effective 'start' which is slightly later than the larger of the two waves reaching the ear. If we were to carry out a similar analysis of what happens at the listener's right ear we would find that the resultant sound wave there would 'start' a little later still. There is, in effect, a TOA difference. This is an important result and perhaps we should re-state it: *for satisfactory stereo imaging the two channels should differ only in amplitude.*

We can now see why the widely spaced microphones used in the 1950s didn't work at all well. Figure 12.4 represents a pair of microphones in front of a band and a long way apart.

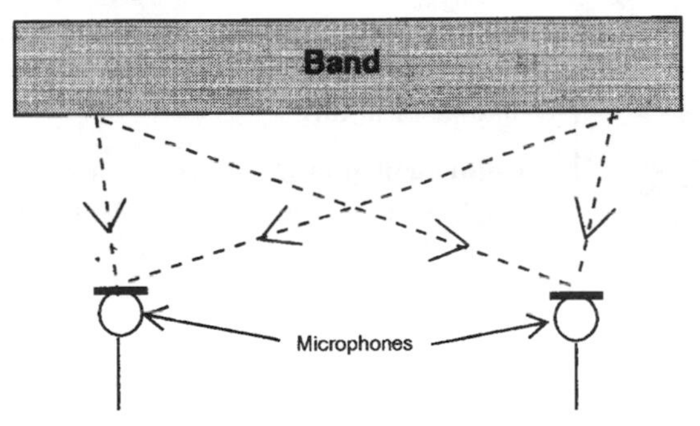

Figure 12.4 Widely-spaced microphones

This means that each microphone receives sound from the left of the band, but with an appreciable time delay for the right microphone, and similarly, a delay in the sound from the band's right for the left microphone. What is going to happen then is that the listener will have a TOA difference from the two loudspeakers, by the process shown in Figure 12.3, and a further TOA difference because of the widely-spaced microphones. The effect is that of an exaggerated stereo effect, so exaggerated in fact that there is a more-or-less empty region in the sound image between the loudspeakers what is known as a hole-in-the-middle. The effect is sometimes called ping-pong stereo. We illustrate the effect in Figure 12.5.

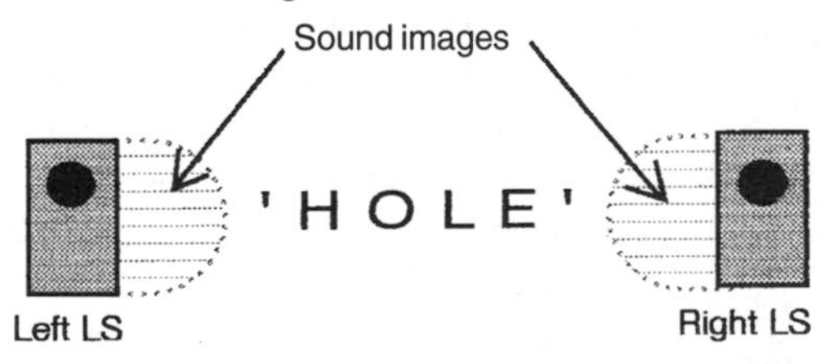

Figure 12.5 Illustrating 'hole-in-the-middle'

Before we leave this section we should point out that some spacing between the microphones may not be unacceptable. For example there are devotees of a system of microphone spacing of about 20 cm, but that is a far cry from the 2 or 3 m which certainly does give a major 'hole'.

12.3 Microphones

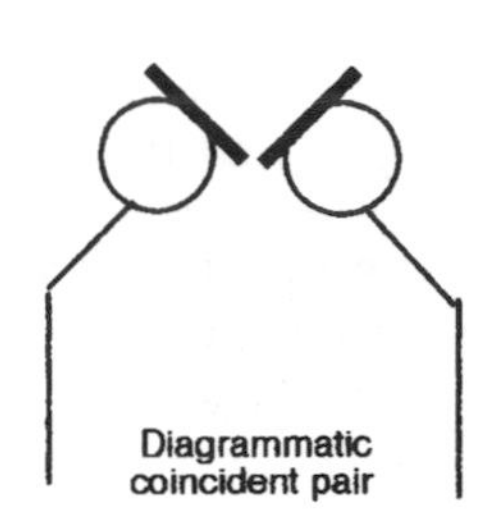

Broadly speaking there are two main microphone techniques. One has two microphones with their diaphragms as close together as possible (as advocated by Blumlein). This arrangement is known as a *coincident pair* arrangement. The other has mono microphones with their outputs positioned electrically. We will take these in turn.

Coincident pairs. Of course it's impossible to have the two diaphragms absolutely coincident in space, but we use the term if the diaphragms are as close together as possible, within a centimetre or two. There are various ways of mounting the two microphones. One is to have them on what is known as a *stereo bar*, which can allow the capsules, assuming that they are electrostatic microphones, to be almost touching. Another arrangement is to make use of specific stereo microphones, which means that the two units are contained within one housing. Such a microphone generally has the (electrostatic) capsules one above the other, the upper one being capable of being rotated. Microphones of this type are usually variable polar diagram ones, so that the individual units can be switched, often remotely, to any useful polar diagram. (The omnidirectional pattern is of no use in producing stereo but it can be helpful in setting up the microphone when it is being rigged.) We shall return to the topic of coincident pairs later in this chapter, but before that we will look at the other main microphone technique.

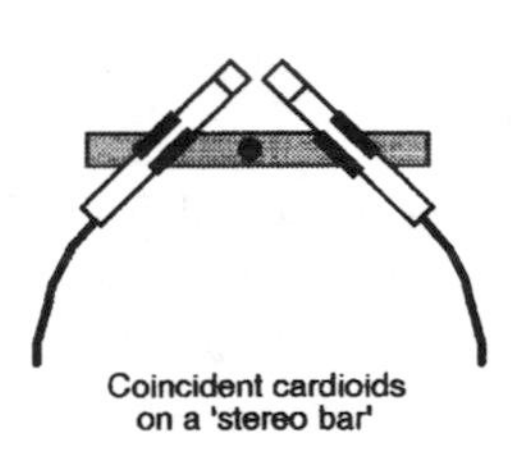

Pan-potted mono microphones. The word *panpot* is short for *panoramic potentiometer*, meaning an electrical potential divider capable of positioning the sound image from a microphone 'panoramically' in the total stereo sound 'picture'. In practice a pan pot control is simply an often quite small knob on a sound mixing desk. There will of course be one for each microphone, so on a sizeable desk there will be many panpots.

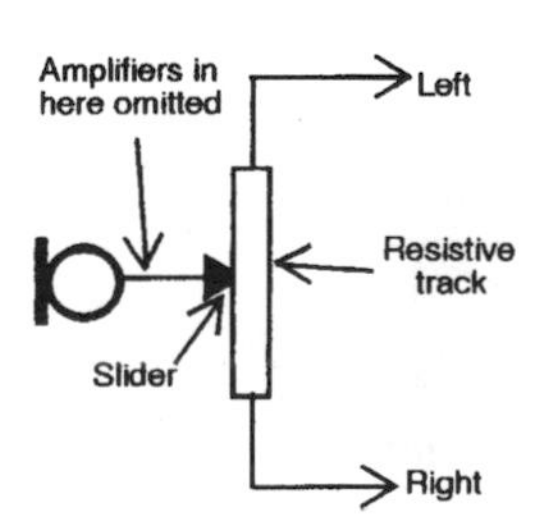

If the slider is at the top of the track in the little diagram then the microphone's output will all go to the left channel; if at the bottom, then to the right channel; and if it's in the centre right then the sound image will be central.

An important thing to notice is that a pan pot only creates amplitude differences in the stereo channels – it does not introduce any time differences, which we've seen is important.

The reader may be wondering when to use coincident pairs and when to use panpots. To some extent this may be a matter of taste: some balancers and producers may prefer one technique to the other. However, and this is perhaps a simplification, there are broad generalizations which can be made.

A coincident pair system can really only be satisfactory when the sound source is *internally balanced.* That means with, say, an orchestra that the relative loudnesses of the various sections are under the control of the conductor and the overall balance is correct in the studio. Any band or orchestra playing to a live audience *without amplification* must be internally balanced. If it isn't then the audience will hear a very unsatisfactory version of what they came to listen to! With an internally balanced source one can think of the microphones as simply eavesdropping on the music.

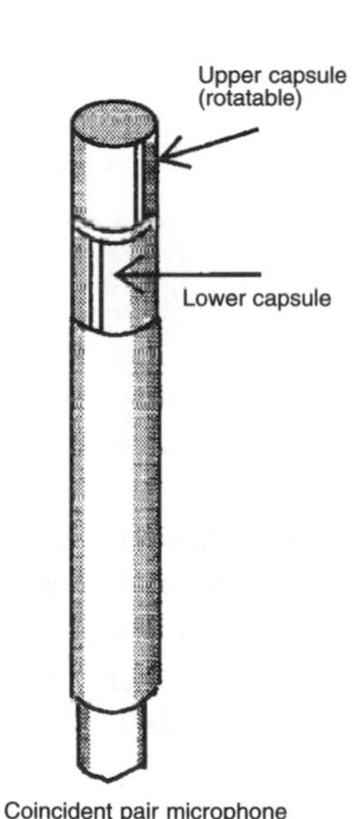

Coincident pair microphone

Most 'popular' music is not internally balanced as it has been written or arranged with the intention of the balance being carried out at a sound mixing desk. A person in the studio will hear a sound which is very unlike that which will eventually come out of loudspeakers. Singers and quiet instruments such as acoustic guitars will probably be drowned by the louder instruments. *Multi-microphone* techniques are then going to be necessary, with, broadly speaking, one microphone for each instrument. To give a stereo result each microphone's output has to go to a pan pot, and the stereo imaging is produced by adjusting these.

What we have just said has, inevitably, some exceptions. To take just one, a coincident pair arrangement may be fine when there is an orchestra in a good acoustic environment, but if the acoustics are poor then the microphones could pick up some undesirable effects and it may be necessary to adopt a multi-microphone approach. And in any case, even when there are good acoustics it is quite usual for a few *spot microphones* – mono microphones – to be used to strengthen possibly weak sections of the orchestra. Some care has to be taken with these to make sure that their pan potted images coincide with the images produced by the main pair. Radio drama is invariably covered with coincident pairs as it would be quite impracticable to try to follow an actor's movements with a pan pot.

12.4 Stereo image widths

With coincident pairs the polar diagrams have a marked effect on the image width, that is the degree of spread of the stereo image between the loudspeakers. Figure 12.6 shows what happens with a pair of figure-of-eight microphones at 90° to each other. The angle of acceptance of such an arrangement is itself 90°, and with the microphones at X the rectangle representing the sound source just fills the 90° as shown in Figure 12.7 (left diagram).

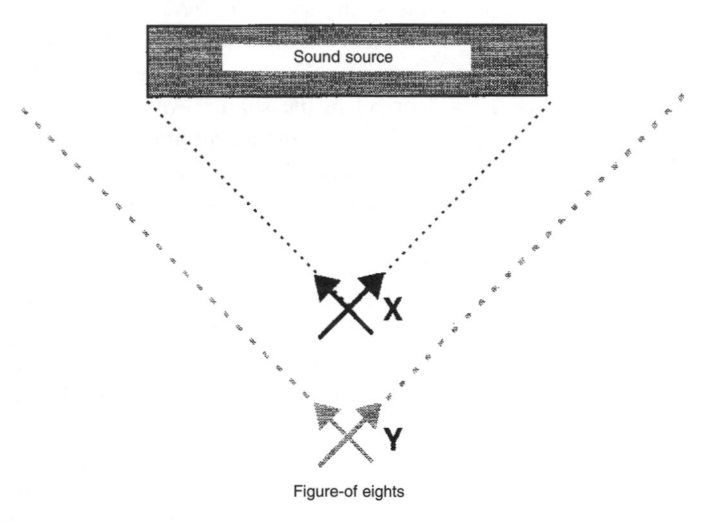

Figure 12.6 Illustrating 'angle of acceptance'

If the microphones were further back, at Y, then the source doesn't fill the angle of acceptance and the resulting sound image is narrower, as in Figure 12.7 (right). (It often helps to think of the microphone pair as being like a camera. If the camera is moved back from a scene any subject in the scene appears smaller.)

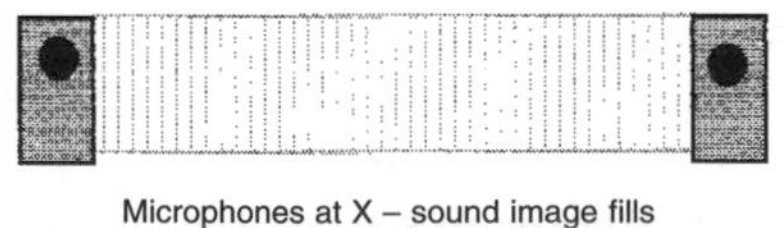

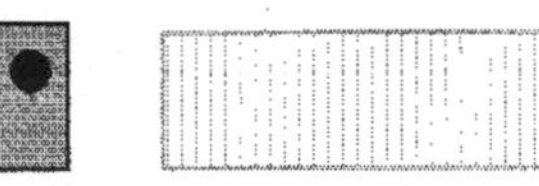

Figure 12.7 Sound image widths

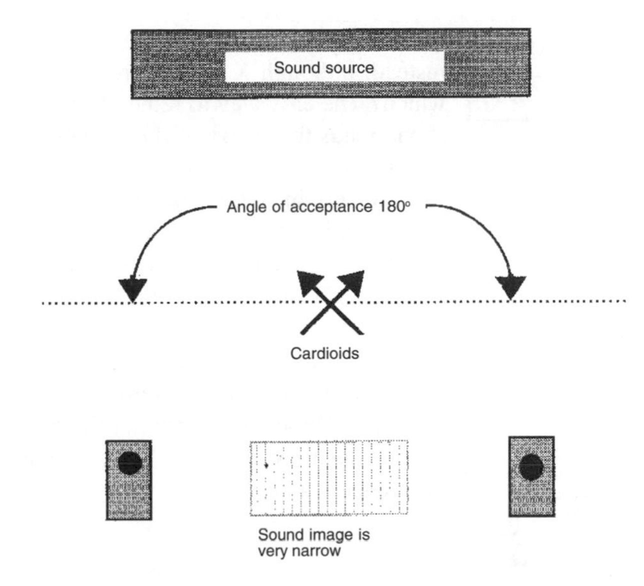

Figure 12.8 Cardioids may produce a narrower image

If the microphones were cardioids instead of figures-of-eight then the result is rather different. The angle of acceptance for two cardioids at 90° is about 180°, so with two of them in the same place, X, as the figures-of-eight we have the situation shown in Figure 12.8.

With the two cardioids the only way that there could be a sound image all the way between the loudspeakers would be for the microphones to be very close to the front of the sound source and this could cause all sorts of balance and perspective problems. The way out of these would probably to use a number of spot microphones. It gets very complicated! Despite this a pair of cardioids can often be found to be useful. Before going on to another topic it is worth mentioning that an angle of 90° between the microphones of a coincident pair is the usual thing. Other angles are sometimes used but there has to be quite a large difference from 90° for image widths to alter significantly. A few degrees on either side of 90° is not normally significant.

12.5 A, B, M and S

In other words, more terminology. So far we have used 'left" and 'right' to refer to the two channels of a stereo system. Indeed 'L' and 'R' are common, especially in domestic equipment. In

ABCDEFGH

the UK professional world the letters 'A' and 'B' are often used instead, although X and Y are common in Europe. Which is which? The easy way to remember that A = L and B = R is that if one reads the letters of the alphabet one goes from A to B in reading from Left to Right!

Also certain colours are used to denote A and B, these being red for A (Left) and green for B (right), which are easy to remember as being the same system used for navigation lights on ships and aircraft. The port (left) lamps are red, the starboard (right) lamps are green. (To complicate things stereo headphones are usually coded with red for the right ear piece!)

There are two other letters associated with important concepts. These are M and S. We speak of 'the M signal' and this means the sum of the A and B signals; the 'S signal' is the difference signal:

$M = A + B$
$S = A - B$

$$M = A + B$$
$$S = A - B$$

(or B – A; it depends which is the greater but it is normally written as S = A – B.)

The significance of M and S is that the M signal is the one which a listener in mono would receive; the S signal can be thought of as carrying the stereo information. This is not strictly true, but it's a good approximation to the truth. (Originally M stood for 'mid' and S for 'side'.) We shall later see that meters for monitoring stereo signals can very usefully show the M and S signals as well as A and B. However in this chapter we are dealing with microphones and mention must then be made of M/S microphones. An important point is that A/B signals can be derived from M/S signals and *vice versa*. After all if:

$$M = A + B$$
$$S = A - B$$

then if we add the two equations we have:

$$M + S = 2A$$

and subtracting gives

$$M - S = 2B$$

We shall meet M and S at intervals in the book. Of immediate interest is the fact that we now have a further stereo microphone configuration.

12.6 M and S microphones

Diagram of an M/S microphone. The single-headed arrow represents, say, a cardioid. The double-headed arrow represents a figure-of-eight microphone

The M and S signals can have their own polar diagrams. If we take a pair of cardioids at 90° to each other we can find the S equivalent by simply adding together the sensitivities at as many angles as we wish, and similarly we can subtract the sensitivities to find the S diagram. If we do this we find that the addition (the M signal) is a not-very-good cardioid it's a cardioid without a zero sensitivity at 180°; the S signal diagram is a figure-of-eight on its side. What this means is that if we have a pair of microphones, one which is similar to a cardioid facing forwards, and a figure-of-eight microphone facing sideways, we have, if we combine their outputs correctly, the equivalent of a pair of cardioids at 90° to each other.

And we find that a pair of figure-of-eights, one facing forwards and one facing sideways gives the M/S equivalent of a pair of figure-of-eights similar to the coincident pair we have looked at earlier.

There are certain advantages in using M/S microphones. One is that, particularly in television work, where a boom operator has to point a microphone fairly accurately, the fact that one of the microphones (the M one) is pointed at the action is helpful. Also, the output of the forward-facing microphone is automatically the mono output. There are, though, one or two problems, none of them too serious. One is that the M and S signals have to be converted into A and B for loudspeaker listening. The circuitry for this is quite simple, as the side diagram shows. Such a circuit can be used both ways converting M and S to A and B, or A and B to M and S. Claims have been made that M/S microphones give better sound qualities than an A/B pair. The possible reason for this is that with a pair of cardioids as an A/B pair the centre line is off the best axis of the microphone and with less than ideal cardioids there might be slightly inferior quality. There's certainly nothing inherent in the M/S microphone principle which makes it superior to the A/B microphone in terms of actual sound quality.

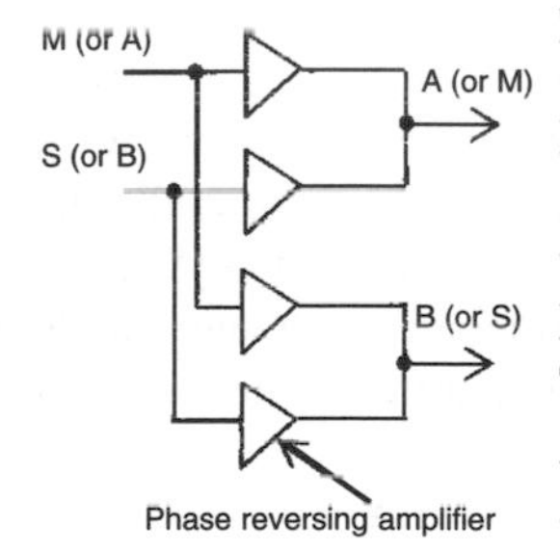

12.7 Compatibility

It's quite obvious that listeners in mono are going to hear something different from listeners in stereo, whether we are talking about radio, television or CDs. The former are not going

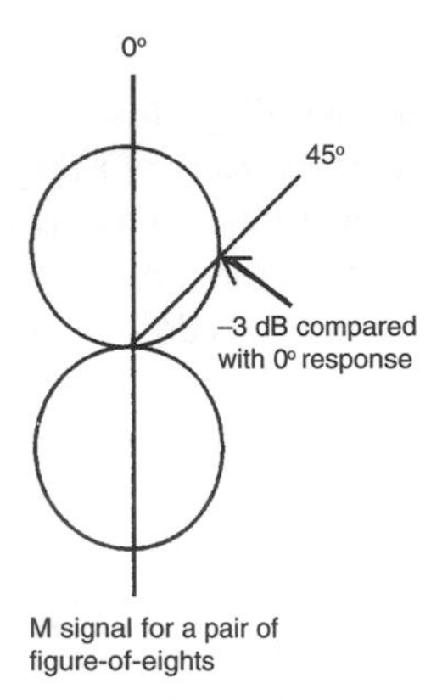

M signal for a pair of figure-of-eights

to have any sense of sound image positioning. But there's more to it than that.

To take an example the M signal for a pair of figure-of-eights is a forward-facing figure-of-eight. A listener in mono will hear a sound output that would be the same as that from a single figure-of-eight microphone. The small diagram shows that sound arriving at such a microphone at 45° off the front will be reduced in level (actually by 3 dB) compared with sound on the 0° axis. Yet listeners in stereo will receive a sound level for that angle of 90° that is actually on the axis of a figure-of-eight microphone! The mono listener is thus going to hear a slightly different balance, with sounds towards the edge of the 'picture' weaker than the central sounds. Whether this is always very serious in practice is another matter. The author has heard at least one recording where the mono version sounded better than the stereo one! Nevertheless, this *compatibility* between mono and stereo is something which one needs to be aware of. When we deal with sound mixing desks we shall see that a useful facility is to be able to switch away from two loudspeakers for stereo monitoring to one to simulate what the mono listener will hear. This makes possible an immediate check on the compatibility of the programme sound.

12.8 Phase

We have mentioned phase earlier in connection with microphones. It's very important indeed in stereo. Normally the signals in the two channels are in phase but differ at any instant in amplitude, as we've shown. In extreme cases the amplitude in one channel is zero, and that means that the sound image is fully to the left or right, depending on which channel does not have a zero amplitude signal. It can happen, though, usually as a result of faulty wiring somewhere, that the phase in one channel is reversed. When that happens one loudspeaker cone is moving out, causing a compression, while the other cone is moving in, not necessarily by the same amount, producing a rare faction. This is a situation which never occurs in real life, or if it does, only momentarily, with the result that the listener's brain cannot make sense of the sound that is entering the ears. It then becomes almost impossible to locate the source of the sound and if one turns one's head the image is likely to move. The sensation is generally quite unpleasant and prolonged exposure can cause a

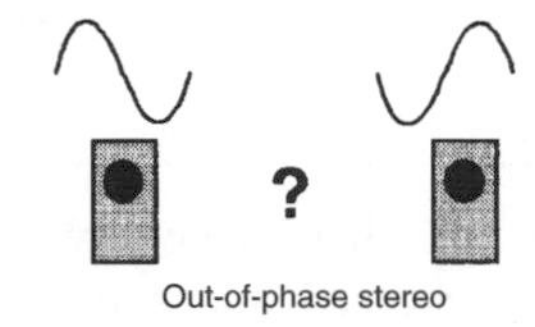

Out-of-phase stereo

headache. There are even reported instances of physical nausea. Consequently out-of-phase stereo must be avoided. Sometimes it isn't easy to be quite sure whether a stereo signal is in phase or not. However, when we come to deal with monitoring arrangements and facilities on sound desks we shall see that there are methods of avoiding an out-of-phase situation.

12.9 Setting up a coincident pair microphone

It is very important that any coincident pair arrangement is correctly set up so that the sound images it produces are where the operator wants them to be. For example they must not be reversed (left to right), and so on. There are some quite simple and logical steps which ensure that everything is satisfactory. Slight variants in the procedure are possible, but the following give a good guide. It needs two people, one at the microphones and one in the control room and able to fade up or down the individual microphones while listening to the result and also observing what happens on the meters. There must be a means of communication, for example by a loudspeaker, from the person in the control room to the person on the studio floor.

Step	Action	Reason
1	Check that both capsules are working.	Obvious!
2	If the capsules are of the variable polar diagram type check that each is set to the same pattern and that the polar diagrams really are what they are supposed to be.	This is easily overlooked but is clearly vital. It is not unknown for a fault to develop and cause a capsule to have the wrong polar diagram.
3	Point both capsules in the same direction. Fade them up one at a time and make adjustments on the mixing desk so that each has the same output on the desk.	Clearly both must have the same sensitivity but this step is going to be important in the phase check (Step 4).
4	Fade up both capsules together. Does the desk output go up or down? If in doubt switch in a phase reverse on the desk.	We went through this in checking the phase of two mono microphones. If the two are in phase then the combined output will go up by 6 dB. If they are out of phase then the output will drop.

Step	Action	Reason
5	Set the capsules to the desired angle (probably 90°). The person in the studio then walks across the sound stage, speaking clearly and identifying his/her position *as seen from the microphones*.	This makes sure that images are not laterally reversed, and it also gives a reasonably good check that the final results will be satisfactory, at least in terms of image positions.

13

Signal processing

By this we mean deliberately altering the character of the audio signal. Many processes are available but we can divide them into two:

1. Frequency correction
2. Dynamic range manipulation.

We will take these in turn.

13.1 Frequency correction

This is actually not a very good term. It suggests that frequencies are corrected, which is not the case at all. What it *does* mean is alterations to the frequency response of a piece of equipment in order to achieve a different balance of the frequencies in a sound. Why we should want to do this will become clearer as we deal with the various methods, of which there are several primary ones.

We will start with *bass cut* which is illustrated in Figure 13.1. We shall look at this in some detail because many of its features apply to other processes.

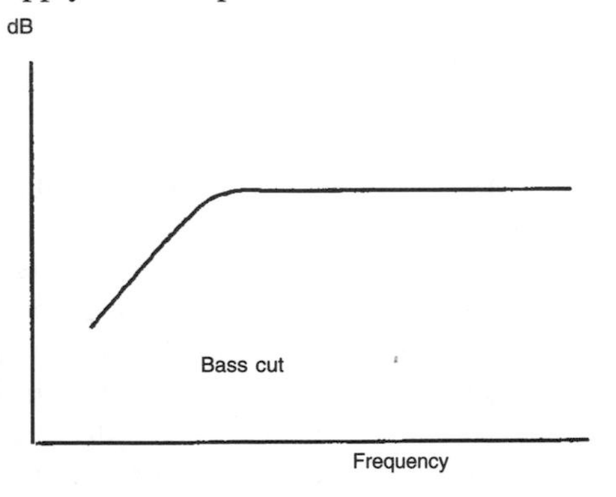

Figure 13.1 Bass cut

It is pretty obvious that somehow the signal level has been reduced at the low frequency end. (In fact quite simple combinations of resistors and capacitors can produce frequency response curves like this.) The reader will notice that we have not put any numbers on either the frequency scale or the vertical 'output' scale. This is because at the moment we don't need to.

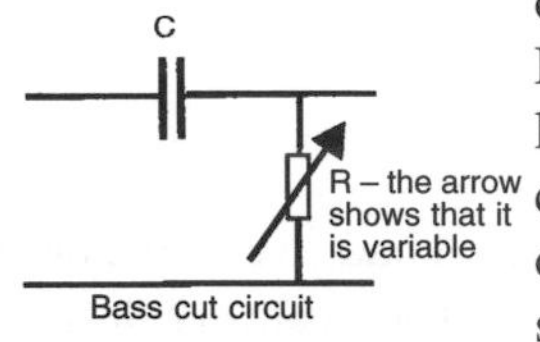

Let us first think about the frequency at which the curve falls off. To begin with this is not a sharp and well-defined frequency. It is standard practice to take the frequency at which the signal level has fallen by 3 dB as a useful criterion in describing the characteristics of the graph. This frequency is called, amongst other terms, the *turnover frequency* or the *3 dB down point*. We shall use the first of these two.

Secondly there is the slope of the descending part. This is quoted as so many *dB per octave*. Basic, simple, curves have a slope of 6 dB/octave, this being quite easy to achieve with few components. We illustrate these terms in Figure 13.2.

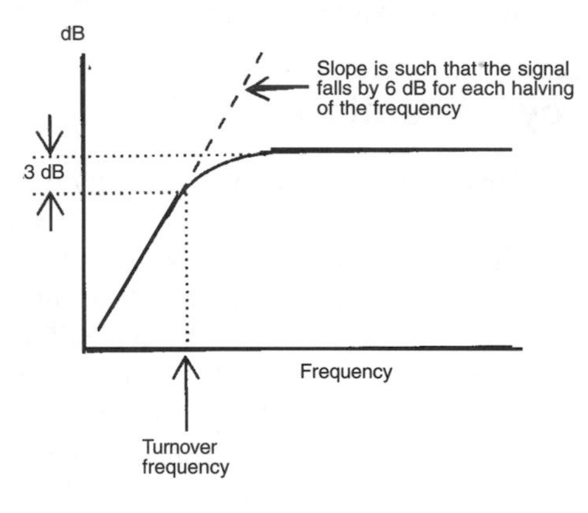

Figure 13.2 Illustrating '6 dB/octave'

A word or two more about turnover frequencies and slope. A slope of 6 dB/octave means that, in this case, the output of the circuit *halves* (halving is equivalent to a 6 dB decrease) for each halving of the frequency. (Of course, with other circuits it's necessary to substitute 'doubling' for 'halving', as appropriate.)

Bass lift and *Top* (or *Treble*) *cut* and *lift* are all shown in Figure 13.3. They are basically similar in having, at least in these simple arrangements, slopes of 6 dB/octave and it is common to have variable turnover frequencies, controlled by a simple rotary knob.

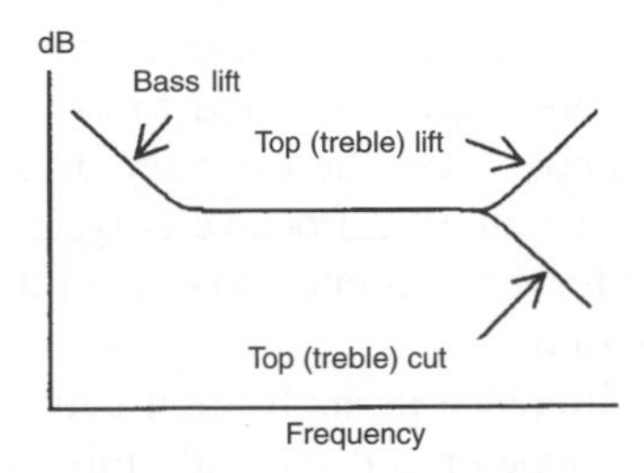

Figure 13.3 Bass cut, top cut and top lift

The reader with a taste for slightly deeper theory might be interested to know that the turnover frequency for a combination of a resistor and capacitor is that for which $R = X_c$.

That is, $R = 1/2\pi fC$

or $f = 1/2\pi CR$

If $R = 1000\ \Omega$ and $C = 0.1$ mF then the turnover frequency f is:

$1/2 \times 3.14 \times 0.1 \times 10^{-6} \times 1000$ Hz $= 1592$ Hz

We can list a few (and only a few!) of the applications of the bass and treble lift/cut circuits:

Circuit	*Typical uses*
Bass cut	1. To reduce 'boominess' in a recording made in a hall with excessive low frequency reverberation 2. To minimize the effects of distant traffic noise
Bass lift	This is generally less useful than bass cut but can sometimes improve things when, for example, the placing of a microphone has been such as to pick up inadequate bass from a musical instrument.
Top cut	1. Reduce sibilance (too many 'sss' sounds in speech) 2. Reduce unwanted high frequency noise such as air-conditioning whistle
Top lift	Like bass lift this is only useful occasionally, but used judiciously it can, for example, 'brighten up' dull speech caused by less than ideal micro phone placing.
Bass cut + top cut	A combination of bass and top cut can be made to give an acceptable simulation of telephone-quality speech.

Besides the fairly gentle circuits we have looked at, a further, sometimes invaluable, set of frequency correction arrangements are what are usually referred to as *filters*. These have much steeper slopes than the ordinary bass and treble lift and cut circuits – about 18 dB/octave is typical, and there are usually two, one for the bass end and one for the high frequency end of the spectrum.

The difference in application between the gentle slope and the steep slope ones is basically this: the gentle slope ones can be used to improve, for a variety of reasons, an audio signal, as we have seen; filters are more for cutting out unwanted sounds.

Figure 13.4 shows a frequency response graph for a bass filter and, on the right, how it may be effective in reducing greatly, in this example, 50 Hz mains hum. A gentle slope filter might have unwanted 'side effects'.

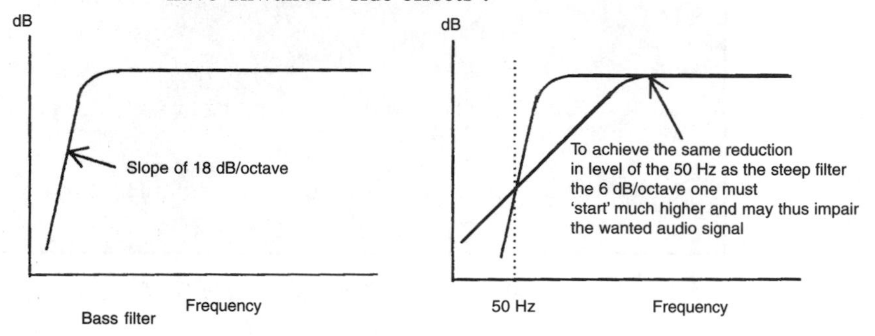

Figure 13.4 The nature of a filter

A further, and perhaps more subtle, frequency correction device provides what is known as *presence*. This, as Figure 13.5 shows, gives a lift or a dip at some frequency which is usually selectable.

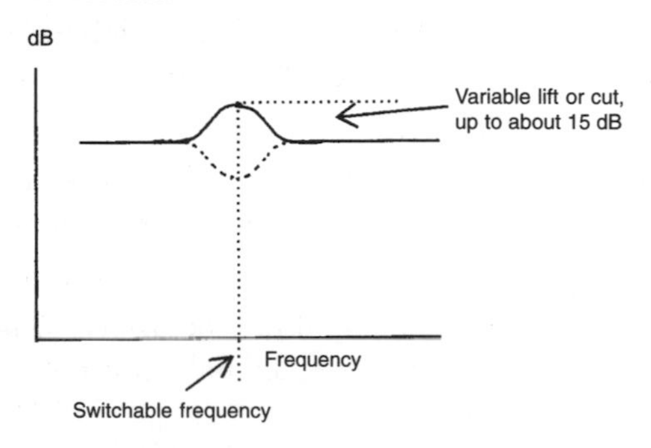

Figure 13.5 Presence

The term 'presence' comes from an observation made back in the late 1950s that a lift of only a few decibels at a frequency of about 2 to 3 kHz had the interesting effect of seeming to bring certain sounds, and in particular voices, nearer. They acquire, as it were, a certain presence. There has never been, to the author's knowledge, a satisfactory explanation of this, but there is no doubt that the effect does occur. A presence control is now a standard facility in any reasonably comprehensive signal processing system, giving either lift or cut of up to about 15 dB at a range of frequencies. The uses are many. A controllable degree of enhancing or weakening of part of the audio signal can be very helpful.

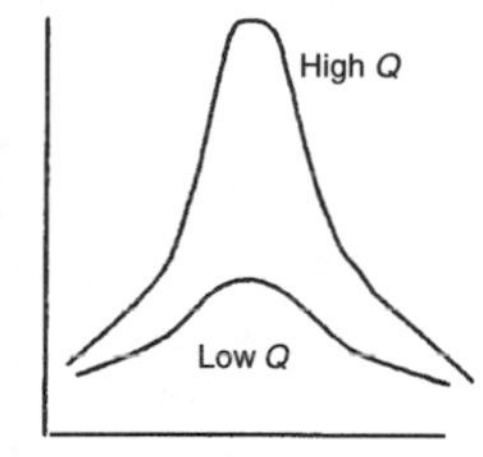

A variant, called for no very good reason a *parametric equalizer* (it does not equalize parameters!), provides presence lift or cut, not at fixed frequencies but over a continuously variable range. This gives great flexibility, especially as it may be possible to vary the width (bandwidth) of the lifted or cut band. This is usually called the Q, from the use of this letter to indicate the bandwidth of electrical resonant circuits. A good illustration of the use of the cut function in a parametric equalizer is in the near-elimination of a particular unwanted noise at one frequency. Suppose this is an intrusive whistle. The best way of reducing it is to set the equalizer to 'lift' and then vary the frequency control until the whistle increases in loudness. If the level control is now turned to 'cut' the whistle will be much reduced.

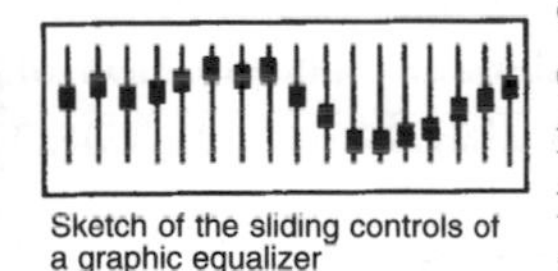
Sketch of the sliding controls of a graphic equalizer

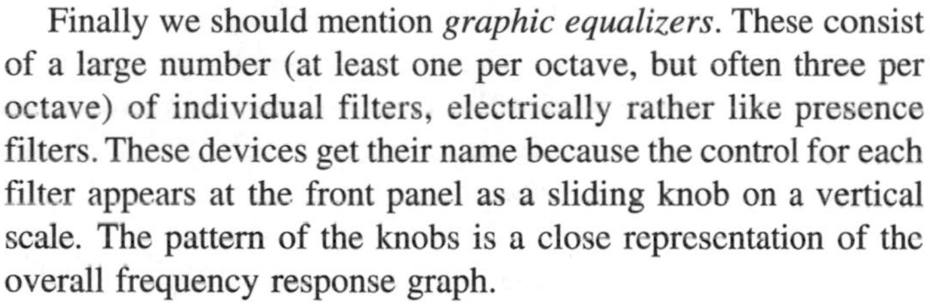
Finally we should mention *graphic equalizers*. These consist of a large number (at least one per octave, but often three per octave) of individual filters, electrically rather like presence filters. These devices get their name because the control for each filter appears at the front panel as a sliding knob on a vertical scale. The pattern of the knobs is a close representation of the overall frequency response graph.

13.2 Dynamic range manipulation

First we had better explain dynamic range. In this book we shall take it to mean the number of decibels between the very loudest and the very quietest levels in the audio signal. This can vary. Probably the greatest dynamic range occurs in symphonic music, where at a point just in front of an orchestra the loudest sounds might peak, momentarily, at 110 dBA, possibly more. The very quietest passages may be only just above the general background noise, the level of which might be 30 dBA at a public concert, and less than that, perhaps 20 dBA, in a recording or broadcasting studio. For this type of music, then, we could speak of the dynamic range as being about 100 dB, if not more.

The range for rock music is likely to be much less; the peak sound level will almost certainly be very high but the quietest parts are still going to be fairly loud. The dynamic range for most rock music is likely to be not much more than 50 dB – at the very most. Less than 10 dB is not unknown!

So much for the dynamic ranges of sound sources. We need now to think about the dynamic ranges which can be handled by transmission systems and recording media. The limits here are, at the upper end, distortion caused by the audio signal voltages going beyond the levels which can be handled by the system and at the lower end a limit set by the inherent noise in the system. We shall go into these limits in more detail in the next chapter.

To take a few examples: the effective dynamic range of a.m. (medium- and long-wave radio) is not much better than about 30 dB! A CD (compact disc) can cope with 90 dB, possibly slightly more, which is just, but only just, about enough for symphonic music. All this shows that in almost any practical broadcast or recording situation there is going to be a need to reduce the dynamic range of the programme material. There are two ways of doing this, one using a skilled operator and the other employing automatic devices. Let us state now that the two are not interchangeable!

We shall later see the principles of *manual control* of signal levels. Here we are going to consider the manipulation of dynamic range by electronic means. The basic device is a *compressor*.

What a compressor does is to monitor the level of the signal and use that to control the level of the signal passing through the unit. The important (and rather complex) item, invariably in the form of a chip, is a *voltage controlled amplifier.* Note by the way that in this case the VCA, although called an amplifier, may not actually be doing any raising of the signal level. In fact much of the time it may be giving an 'amplification' of less than unity. Figure 13.6 shows the arrangement in diagrammatic form.

A VCA is special in that its 'amplification' is controlled by a d.c. voltage. In Figure 13.6 the input audio signal is measured and a d.c. signal is derived from it which then goes into the

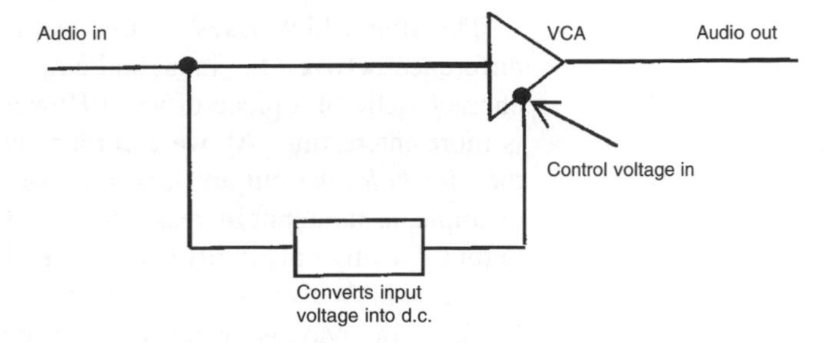

Figure 13.6 Basic diagram of a compressor

VCA. (The loop doing this is called the *side chain.*) If the input level rises above a certain pre-determined value then the side chain voltage causes the amplification to be reduced so that the output doesn't rise as quickly as it would otherwise have done.

The action is illustrated in Figure 13.7. The two axes are both marked in decibels, the horizontal one showing the level of the input to the compressor, the vertical scale showing the output level. (The decibel scales we have used here are not what one would normally find in graphs of a compressor's action. They will always be in decibels but the range of numbers is going to be different. This is a matter we shall deal with later. At the moment we are simply concerned with the basic behaviour of a compressor.)

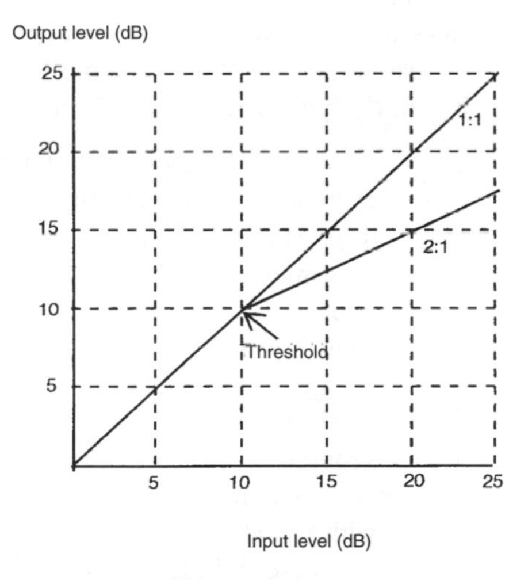

Figure 13.7 The action of a compressor

The line which rises at 45°, labelled 1:1, represents no difference between the input and output levels. It would be the characteristic of a piece of wire! However the line marked 2:1 is more interesting. Above a certain input signal level, called the *threshold*, the output rises more slowly than the input. For example, as the input increases by 10 dB (from 10 to 20 dB) the output rises by only 5 dB (from 10 to 15 dB). There is thus a reduction in the dynamic range of the input signal. The threshold level, by the way, is switchable over quite a wide range. Also the *compression ratio*, which here is 2:1, is also adjustable.

At this point we must give a definition of 'compression ratio'. It is the ratio:

$$\frac{\text{Change in input level (in dB)}}{\text{Corresponding change in output level (in dB)}}$$

above the threshold

The last bit, 'above the threshold', is a very important part of the definition.

Typical compression ratios on a professional compressor might be 1:1, 2:1, 3:1, 5:1 and 10:1. The 1:1 setting may not seem helpful but in fact it is because it is equivalent to the compressor being switched out of circuit – and with many audio processing devices their effect has to be judged by ear and often by comparison with the situation without the device.

Before we go further let us see a few uses for a compressor. In general it's reasonable to say that the most common use is to increase the apparent loudness of a sound. However the following is a fairly typical selection of other applications:

1. Loud sound effects in a drama. It may be difficult or even impossible manually to operate a fader fast enough to prevent overload and consequent distortion.
2. As a means of holding the output or group of microphones at a more-or-less steady level to prevent their sources from becoming too loud in relation to another source. A compressor might well be used to control the level of a vocal backing group.
3. Recordings of news events, particularly in the open air, are situations where it may be difficult for a sound operator with a small recorder to hold a microphone and at the same time use a fader to control unpredictable levels. A compressor built in to the recorder can be invaluable (as long as it can be switched out when not wanted).

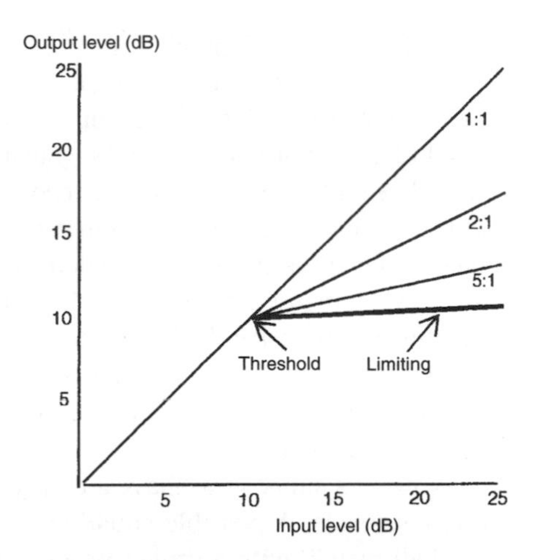

Figure 13.8 Some compression ratios and limiting

We have yet to mention *limiting*. This is sometimes spoken of as if it were something entirely different from compression. It isn't. Limiting is simply an extreme case of compression, where the compression ratio is very high. Ideally it would be infinite – in practice it may be 20:1 or more. Figure 13.8 shows compression ratios of 1:1, 2:1, 5:1 and limiting.

There is a minor complication here, but only a complication if one is not expecting it, in that some *limiter/compressors* (units combining both functions, possibly independently) automatically raise the threshold by 8 dB for limiting. 8 dB is not a randomly chosen number. We shall see later why, at the moment it must be enough say that average level of most programme sound is about 8 dB below the maximum permissible level. Thus a threshold shift of 8 dB would mean that limiting would prevent any tendency to exceed the maximum level.

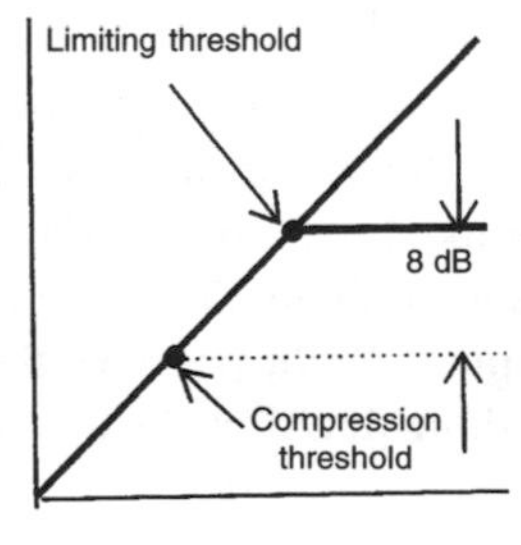

There is a most important aspect of limiters and compressors which we must mention. It is concerned with the time it takes the VCA to return to normal (1:1) operation after there has been a reduction in level. At first sight this might not seem very significant, but in fact it is. We are referring to what is known variously as *recovery time, decay time* or *release time*. We shall use the first of these, 'recovery time', as being the best description of the effect. The best way of explaining why recovery time is important is to give an example or two.

Suppose the VCA returned to normal operation instantly after there had been a gain reduction, so that the recovery time was zero. Take classical music: after a loud passage with marked level reduction the following quiet music would, on reproduction, be only a little quieter. The contrast between loud and quiet would have been lost, or at least much reduced, and the result would be uninteresting. A longer recovery time, several seconds at least, would mean that the quiet passage following the loud one would initially be much quieter but would slowly rise in level. To some extent therefore the contrast between loud and quiet would be partially maintained. (With classical music it is, if possible, much better to have skilled manual control rather than rely upon automatic devices. These are quick in operation but not at all intelligent!)

Now let's suppose we have a relatively long recovery time – a few seconds. A possible situation is with a commentary at a football match, with a limiter or a compressor in the output of the system to guard against a sudden and excited raising of the commentator's voice or of the crowd noise. Suppose there is a big increase in the crowd noise. The device reduces the gain which only slowly comes back to normal. This means the commentary is too quiet for a few seconds, gradually coming back to normal level – until the next goal is scored! A short recovery time would be better, although there is then an unnatural effect of the louder noise 'punching holes' in the quieter one. This is often called *gain pumping*, which is actually a fairly good description. There may be a way round the whole problem – see the diagram on the next page.

These two examples show that recovery time is important and most limiter/compressors have a control for selecting it. A typical range is from 100 ms to 3.5 s. There is often a further position called 'auto'. This varies the recovery time according to the duration and level of the sound which caused gain reduction to occur.

Many professional units have yet another control – *attack time*. This operates on the speed with which the VCA reduces the level after the threshold has been reached. It might seem as if this should be as short as possible, but in fact it can happen that a very quick attack time of a few milliseconds affects the important starting transients which we mentioned in an earlier chapter, thus altering the quality of a sound. On the other hand a slow attack time – 100 ms for example – might preserve the starting transients but with a risk of short duration distortion

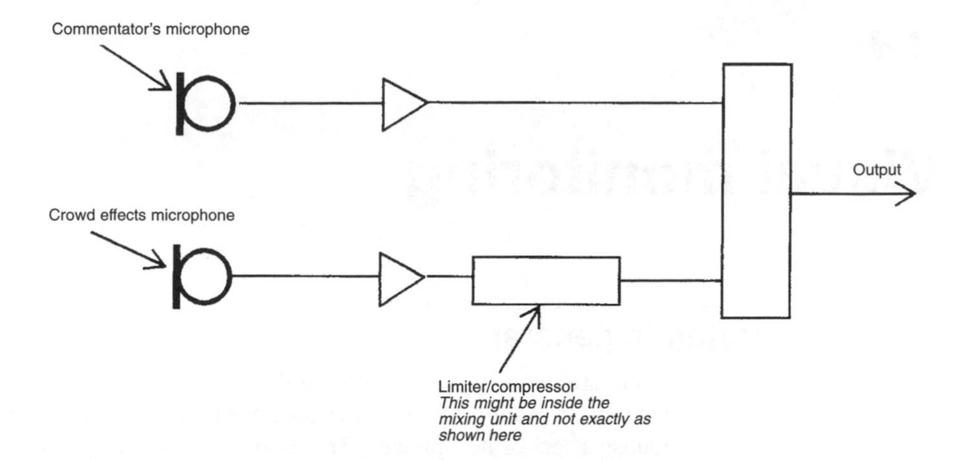

cause by an overload at the start of the sound. It is a matter of careful judgement and compromise.

Finally, a mention of *noise gates*. These are very similar in basic concept to compressors except that they increase the dynamic range. A noise gate is chosen to operate at low signal levels and is set so that unwanted studio noise, for instance, is below the operating point and thus has a much reduced level in the output. The dynamic range is 'stretched' at these levels. With carefully chosen operating settings noise gates can be very useful, but like all automatic devices they have their limitations – if the programme sound drops to the noise gate threshold it disappears because the device cannot distinguish between wanted and unwanted sounds. It can only interpret levels.

The diagram above suggests how the noise from a football crowd (or similar) can be controlled automatically without gain pumping on the commentary.

14

Visual monitoring

14.1 Monitoring in general

There are two types of monitoring that we are concerned with. One, *aural monitoring*, uses the operator's ears with the aid of loudspeakers or headphones. The human ears are very good at detecting distortion and things like *balance* – the relative loudnesses of sources – and also perspective. No electronic device has yet been made to replace the ear in these matters. However, the ear is very poor indeed at assessing *absolute* loudness. And with that goes its inability to decide when signal levels are too low or so high that there is a risk of distortion. (Actual distortion is, of course, audible but we'd like to have a warning of when the signal level is approaching the distortion region.) For this we have to rely on *visual monitoring*. In simple terms that means things like meters, and in this chapter we shall look at the most important types available.

First, though, we will follow up some of the things outlined in the last chapter.

14.2 Dynamic ranges in technical terms

Any audio signal in its electrical form has two limitations in respect of its amplitude (the same goes for video signals). At the lower end there is the noise inherent in the system and at the upper end there is the level which the system just cannot handle properly without distortion. Figure 14.1 illustrates this.

Taking the upper limit first, any amplifier will only deal with electrical signals which are in its proper range. In the case of an amplifier which is used to raise the very weak voltages (millivolts or less for much of the time) from a microphone up to a convenient level it is likely that an input voltage of a few hundred millivolts would result in a distorted signal.

The lower limit we have referred to is determined by the inherent noise in the system. In the last chapter we pointed out

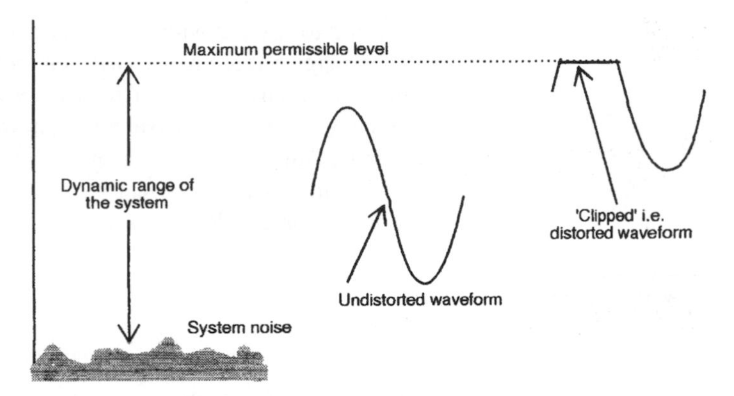

Figure 14.1 Dynamic range

that an electric current consists of a flow of individual particles – electrons – and not a continuous flow as if of a fluid. Ultimately the electron flow can manifest itself as a hiss – with good modern audio equipment this is at a very low level indeed, but it's there. (This noise is related amongst other things to temperature and is thus called *thermal* noise. The higher the temperature the greater the noise. In some radio telescopes designed to pick up the incredibly weak radio signals from, say, distant galaxies the first part of the radio receiver is cooled in liquid helium, to a temperature of about −269°C, which is only four degrees above the absolute zero of temperature.)

If the signal is to be recorded magnetic tape has an upper signal limit which is the *saturation* level because all magnetic substances reach a stage where they are fully magnetized. There is a lower noise level caused by the fact that the magnetic particles on the tape are sufficiently large for each one to contribute its own minute spike of voltage, although ground down in the manufacturing process to a very fine powder. With modern tapes and specialized techniques this *tape hiss* is small. Nevertheless the reader will surely have noticed a slight hiss with cassettes when the non-magnetic leader has passed the heads and the recordable section starts. (Digital recording effectively eliminates the effects of tape hiss.)

So we have, then, an upper and a lower limit to the dynamic range in any system which amplify, record or distribute audio signals. The important thing is to make sure that the signal stays between these limits, and this is where *visual monitoring* comes in. Before we look at the available types of meter there is one very important aspect to consider, namely what kind of scale are we going to use? Obviously we shall be dealing with decibels for the range but the reader will remember that the decibel is a unit of comparison. With hearing we took as the zero, or reference, a sound pressure which was approximately that corresponding the average ear at its most sensitive. Now, though, we're dealing with electrical signals and the same zero would not be convenient. We need to introduce *zero level*.

14.3 Zero level

We are going to establish an electrical voltage which has a dual function.

Zero level:
0.775 V

First we need something as the zero for the decibel scale. Voltages above this will be given a plus (positive) sign, those below it a minus (negative) sign. The voltage which, for mainly historical reasons, has been universally adopted is 0.775 V. This is a curious sort of number to choose and had the whole business been started today it's likely that 1 V would have been adopted. The interested reader might care to read the explanation below.

Originally one milliwatt (1 mW) of power was taken as a standard for audio signals. This seems to have been chosen because it was well within the capabilities of modest amplifiers, and also the headphones of the day, if plugged into the circuit, gave an adequate loudness for checking the signal. Further, there was a standard impedance of 600 W, and this came about because it was the kind of impedance encountered with a pair of long telephone wires spaced according to the then standard system. With various technological developments the reasons for using 600 W faded away, but the convention of using it persisted. Also the 1 mW was replaced by simple voltages – after all, it's easy to measure voltages but much less easy to measure power.

If we want to replace 1 mW in 600 Ω by the equivalent voltage we use $P = V^2/R$, so that:

$$0.001 = V^2/600$$
$$V^2 = 600 \times 0.001 = 0.6$$
$$V = \sqrt{0.6}$$
$$= 0.775$$

We can now define the zero level as being an audio signal voltage of 0.775 V. (Strictly we should say 'an r.m.s. voltage'.) It then becomes possible to draw up a table (below) of signal levels in decibels relative to zero level.

We said above that this zero level was going to have a dual function. One, we've seen is as a zero for a convenient decibel scale. The second one is as a standard signal which can be used for *line-up* purposes. We'll explain that it's very important indeed that all audio equipment in a studio is compatible in terms of the signal levels going to it and from it. This is even more important in a broadcast chain. For example, the levels coming out of a mixing desk must be appropriate for the recording equipment; the signals from a remote studio must be at a level which is acceptable at other points in the network, and so on.

Level	*Voltage*	*Percent of 0.775 V*
+8	1.95	250
+4	1.23	160
0	0.775	100
–4	0.489	63
–8	0.309	40
–12	0.195	25

Line-up tone: zero level at approx 1 kHz

The process of carrying out the necessary checks is part of a *line-up procedure*. Zero level is used almost universally as the correct voltage. *Line-up tone* is an audio signal of about 1 kHz at 0.775 V. The exact frequency is not critical; the level is. We shall meet line-up tone again.

Now to return to the actual measuring devices. There are two primary kinds of meter for the job and we will look at them in turn. They are known by initials: the *VU (volume unit) meter* and the *PPM (peak programme meter).*

14.4 The VU meter

A VU meter is really only a voltmeter with a special scale. The latter can vary somewhat but the more usual versions have a decibel scale with a red section towards the right, this ostensibly showing the potential overload region. We say 'ostensibly' because there are wide variations in the way these meters are used. For example, in some kinds of equipment, some tape recorders for instance, the user is advised to allow the pointer to go occasionally into the red region. Also the correct pointer

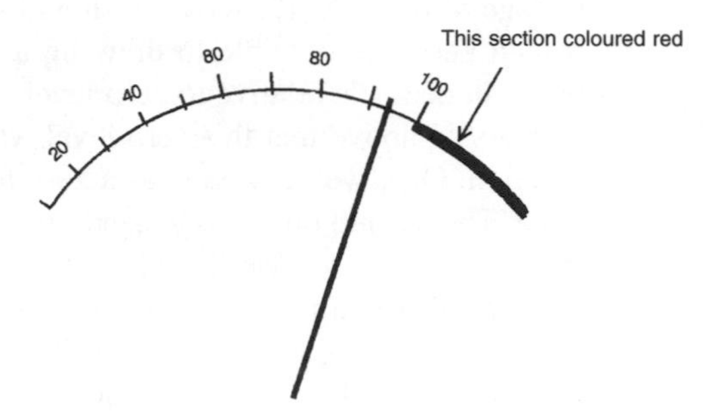

Figure 14.2 Scale of a VU meter

position for line-up tone ought to be 8 dB below the '100%' position but again there can be some variation. The main disadvantage is that the meter reads average voltages, and it does not show where the signal level is in relation to the overload region.

Despite these drawbacks VU meters have their uses. They are relatively inexpensive, they can be used with reasonable accuracy for steady signals (line-up purposes), and being fairly cheap they can be useful indicators simply of the presence of audio signals. An example of the last point is in some types of multi-track tape recorders. A 24-track machine, say, can have 24 VU meters.

14.5 The PPM

The PPM meter differs from the VU meter in almost every way possible. Its most important difference is that it indicates, reasonably accurately, the actual peak signal levels (hence its name). This means that as signal levels increase there is warning of an impending overload and hence distortion. This is made easier by the fact that the pointer has a very rapid rise time but a slow fall-back.

Secondly, as Figure 14.3 shows, the scale and pointer are white on a black background and this makes it easier to read the meter, especially if there is poor lighting. Also there is a minimum of graduations and numbers, and this also makes reading easier.

The third feature is that the scale is linear in decibels (although it may not appear so at first sight!). Notice (a) that the important

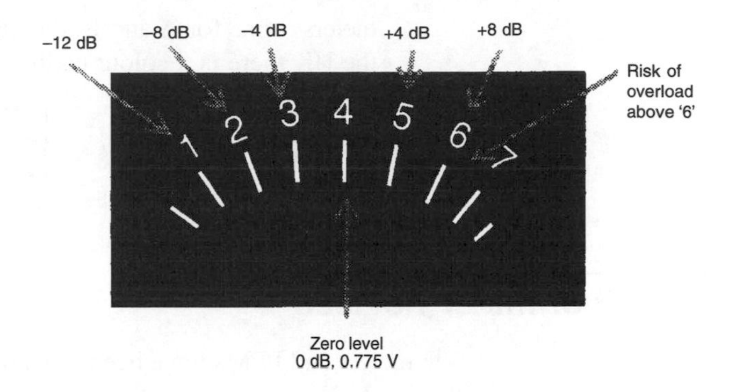

Figure 14.3 PPM (Peak Programme Meter)

reading for zero level is '4' in the middle of the scale and (b) that the scale divisions are 4 dB apart, except at the extremes of the range. This might seem a very coarse scale but with very little practice it is possible to read programme peaks to within about 1 dB. With steady tones about half a decibel can be read.

The table below summarizes the main features of these two instruments.

	VU meter	*PPM*
Indication of peak levels	No	Yes
Scale linearity (dBs)	Poor	Good
Easy to read	Not very	Yes
Steady tone readings	Good	Good
Cost	Low	High

Additional points are:

1. It is sometimes claimed that the VU meter gives a better indication of loudness than a PPM. There is little evidence to support that view. The fact is that neither meter does very well in this respect, for the simple reason that neither makes any allowance for the curious frequency response of the ear.
2. A double PPM makes a very useful meter for stereo monitoring. There are two pointers with coaxial shafts, rather like the hour and minute hands of a watch. Then one pointer can be used for the A signal and the other for the B signal, *or* one pointer indicates M and the other indicates S. It is usual on mixing desks to have two stereo

meters – one for A and B, the other for M and S and in the UK there is a colour coding system:

Red pointer	Left
Green pointer	Right
White pointer	M
Yellow pointer	S

14.5 Other meter devices

VU meters and PPMs have been around for a long time – since the 1930s. In recent years other devices have become available, made possible by advances in technology. There are broadly two types:

1. *LED* (*light emitting diodes*). Typically these small light sources are arranged in a vertical (sometimes horizontal) columns and appear as a line of green lights, the colour being changed to red above the 'overload' region. There is usually associated circuitry that gives a quasi-peak-reading display – the highest LED stays lit for a second or two after the peak has passed. (This is sometimes called a bouncing ball display.)

   LED displays differ in their resolution: on domestic tape recorders, for instance, the individual LEDs may be 4 dB apart, reducing to perhaps 1 or 2 dB at the top end of the range. On professional equipment the resolution is better, say 2 dB over a wide range. The advantage of LED displays is that they take up very little panel space and are easily observed. The accuracy is never as good as a PPM and for line-up purposes a simple VU meter is a useful addition.
2. *Plasma displays*. These are much more complex devices and are often to be found on large sound mixers. They appear as a column of light, changing like LED displays to red above a certain level. The column of light in a typical unit is a centimetre or so wide and perhaps 10 cm long. The resolution is good and an accuracy of 1 dB or better is achievable. On some major sound mixers a long line of plasma displays can be used to indicate levels on each microphone channel, switchable to a VU or PPM response.

15

Analogue tape recording

15.1 Basics of the process

The basic idea of tape recording is simple enough: the audio signal flows in a coil forming part of what is essentially a small electromagnet (the *record head*). The varying flux from this creates a pattern of magnetization in the small particles of magnetizable material on the tape. Then, at the replay stage, the tape passes an electromagnet similar to the one used for recording and the variations in the magnetic field round the tape induce a small e.m.f. in the coil of the head. Ideally this e.m.f. should be a replica of the original audio signal, and in practice should be very close to the original. The rudiments of the process are shown in Figure 15.1.

We will look at the components shown above in a little more detail.

First, *the tape*. This is made of a thin but strong plastic base to which are strongly bonded very finely ground particles of a suitable magnetic material (an oxide of iron). This material has to have several important properties. First it should be capable of being strongly magnetized and hold that magnetization for as long as the recording is needed – which could be many years. Secondly, it must be possible to demagnetize (*erase*) the tape when there is a need to over-record. Third, the magnetic material should not be abrasive so that it wears out the heads as it passes in contact with them.

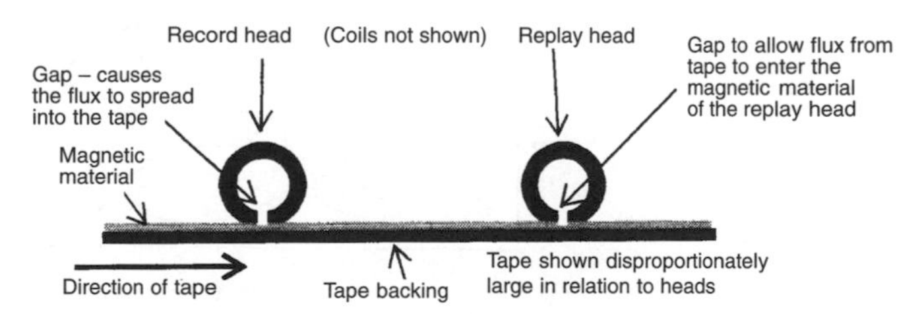

Figure 15.1 Rudiments of magnetic recording

All these requirements, and there are others, have presented manufacturers with serious problems which, over the years, have been largely solved so that high quality recordings can be made.

Now the heads. In fact there are three, not two as shown in the diagram. The third head, which is actually the first head the tape meets, being used to erase the tape before it reaches the record head. These have to be of a material that does not retain any magnetic effects, so that, for example, the flux in the record head follows accurately the current in the coil, falling to zero when the current drops to zero. The sizes of the gaps are important: that for the record head must be relatively large (but even so only about 20 mm (a fiftieth of a millimetre) so that the flux can spread easily into the tape.

The replay head has a much smaller gap, often less than half the size of the record head gap, so that it can sample accurately each part of the recorded signal. The erase head, which is supplied with an a. c.t having a frequency of typically 100 kHz or more, has a large gap of about 100 μm, again so that the high frequency flux can spread fully into the tape. The process is a little more complicated than this and the reader who wishes to find out more of the details should consult some of the books listed in Chapter 23.

We have glossed over a very important part of the recording process. Unfortunately magnetic materials don't have what is called a *linear transfer characteristic*. That means that if we use, for example, a sine-wave magnetizing force then the resulting magnetization of the material is not sine-wave shaped. The problem is solved with the use of *bias* – an a.c. signal mixed with the audio signal to be recorded. This has a high frequency, usually 100 to 200 kHz. The exact frequency is not critical but the level is. Figure 15.2 shows how the bias level affects tape noise and distortion.

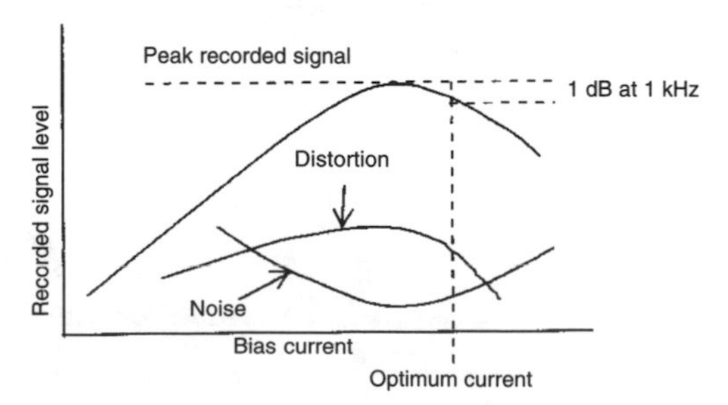

Figure 15.2 How bias level affects noise and distortion

The optimum bias current is *not* that for which the recorded signal level is greatest, nor the distortion least, but at a compromise setting. Fortunately analogue tape machines today are very stable and the bias usually needs to be set only occasionally, although where these machines are used in mobile work they should be checked regularly, as vibration when being moved may disturb the control settings. (At one time tape machines needed to have their bias set every time they were used!) As we shall later see, bias setting is not a problem with digital recorders.

15.2 Moving the tape

This is not as easy as it might seem. The list below gives the most important requirements:

1. The tape must travel past the heads at a constant and accurate speed. Professional machines operate mostly at 38 cm/s (15 inch/s) but 19 and 9.5 cm/s are sometimes used. This means that, taking the standard speed, the tape should travel at 38.000 cm/s at the start and at the end of the reel. This need for an accurate and constant speed is perhaps a little more important than might appear at first sight. To begin with, in the professional world, and especially in broadcasting, it's important that a recorded programme due to fit, say, a 30 minute slot, should run for perhaps 29 minutes 45 seconds (to allow for a 14 second announcement). A 10 second over-run could be very awkward; an under-run would be less serious but could still be embarrassing. The tape speed, then, ought to be such that over about 30 minutes the timing error ought to be less than 3 or 4 seconds.

   Not only that but it may be necessary to do a retake of part of a musical item at the end of the tape. If this is to be edited into the early part of the tape there may be an obvious and unpleasant pitch change if the tape speed towards the end of the reel has drifted slightly from what it was earlier.
2. Not only must the long-term speed accuracy be good but the short-term constancy is also important. It would be possible to have rapid but small speed fluctuations and still have the overall speed within tolerance. Speed fluctuations of more than about one per second are

known as *flutter*; around one per 2 or 3 second fluctuations are called *wow*. Both are unpleasant.

3. The tape must be kept at the right tension. If it is too slack then it may not keep in close contact with the heads. Too tight and there is a risk of the tape stretching.
4. The tape must pass the head gaps at exactly the right angle. If it is not at precisely 90° to the head gap then there can be a reduction in the high frequency response. There is then a form of *azimuth error*, which we shall mention later.

All these things mean that the *tape transport system*, as it is called, has to be a piece of precise engineering and often represents quite a large part of the cost of a high quality machine. (An advantage of digital recorders, as we shall see in a later chapter, is that some of the requirements of the tape transport are less important.)

Figure 15.3 shows, in slightly simplified form, the essential features of a *tape deck*, with the heads and the transport system.

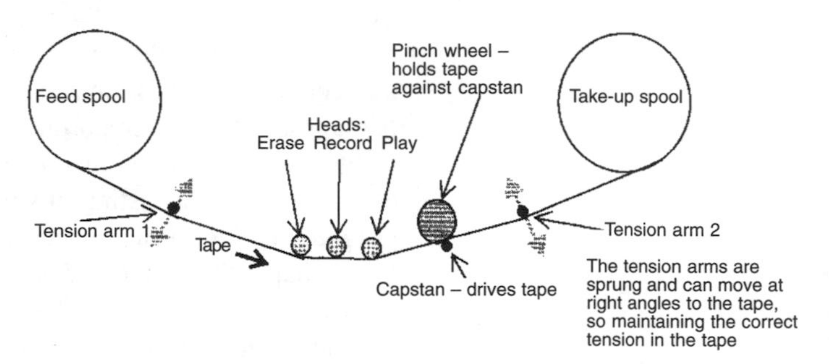

Figure 15.3 Tape transport system

15.3 Azimuth

We said earlier that the replay head gap had to be small to sample accurately all parts of the recorded signal. Put another way, if the gap were, for instance, equal to the *recorded wavelength* at a particular frequency, that is, the length on the tape of one complete cycle of the audio signal, then there would be no output at that frequency. Let us explain this a little more.

Take a tape speed of 38 cm/s, and an audio frequency of 10 kHz. The length on the tape of one complete cycle of this

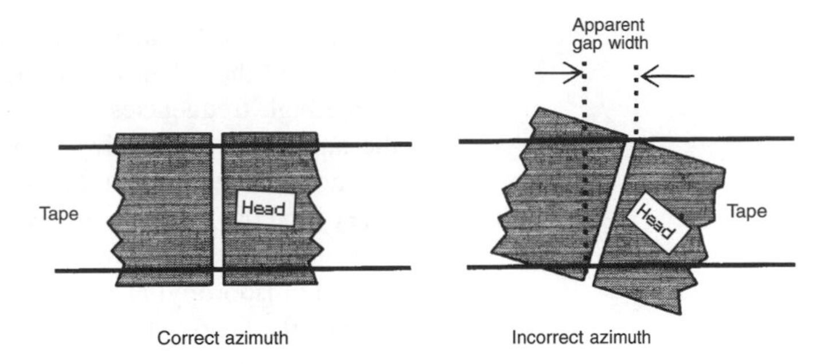

Figure 15.4 Correct and incorrect azimuth

frequency would be 38/10 000 cm, or about 0.04 mm. A gap width of this size would contain one complete wavelength and would therefore not be able to detect any flux difference. The result would be no output at that frequency. (A typical replay head gap width is about 0.01 mm – a quarter of a recorded wavelength at 10 kHz.)

Now suppose the heads of a machine had become out of alignment – through mechanical vibration when being transported for example – so that the gap is no longer exactly at right angles to the tape. This is what is called an *azimuth error* and, as Figure 15.4 shows, is equivalent to a wider than normal gap.

An azimuth error, then, causes a reduction of the audio high frequencies – an unintended top cut! Correction of the azimuth is a rather skilled business. It involves recording and/or replaying a high frequency signal, typically about 10 or 15 kHz, and mechanically adjusting the head for maximum output.

15.4 Tape errors

We have already mentioned errors that can occur with analogue tape. Here are a few more to add to the list, with ways of either correcting them or making them less serious:

1. *Dirty heads or tape path.* It could be said that cleanliness is next to high fidelity! The trouble is that the oxide coating on tape, although very strongly bonded to the backing, can with time and extensive use, leave a very thin deposit of oxide on the heads. There are other possible ways of unwanted deposits. Wax pencils sometimes used to mark the tape when editing can be a major source of trouble. The effect of this contamination is to make it

impossible for the tape to lie as close to the gap as it should and the effect is yet another cause of reduction of the high frequencies. The heads, and other parts of the tape path should always be kept clean. It is good practice to wipe these parts with a soft, lint-free cloth every time the machine is used. With cassette recorders a recommended cleaning method is to use a cotton bud soaked in isopropyl alcohol. (The reader who has never cleaned the heads of his or her cassette recorder should try this. Almost certainly there will be an alarming amount of the brown oxide on the cotton bud afterwards!)

2. *Tape drop-outs*. Modern tapes, even the relatively low-cost ones used for domestic cassette recorders, are made to a very high standard but even so it is possible for there to be small areas where the oxide is too thin or even non-existent. These are known as 'drop-outs', which describes the effect to the listener. They are, as we have said, extremely rare with tapes made by reputable manufacturers but it is best not to rely too much on this. Obviously the user can do nothing to put right a drop-out – all that can be done is to ensure that one is aware as soon as possible of the existence of one. We shall see shortly that *off tape monitoring* provides the necessary checks.

3. *Saturation of the tape*. We made a reference to this when we were dealing with dynamic range. It is simply that an excessively high signal level can cause all the magnetic particles on the tape to be fully magnetized so that any further increase in the signal has no additional effect. We say that the tape is saturated. The answer is clearly not to let signal levels get so high that this happens. A careful check on *input* levels is obviously very important, but a further check is to use off-tape monitoring.

This means that the monitoring meters and loudspeakers are fed with the output of the replay head, not what is being fed to the record head. There is of course a slight time delay between the in-going signal and the monitored one, but this is small. On a full-size professional machine it may be less than one tenth of a second. The point is that by listening to what has been recorded any drop-outs or distortion through saturation are detected almost instantly. What happens next depends

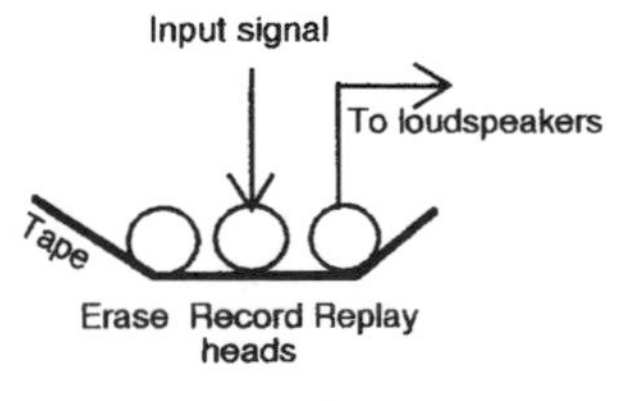

Off tape monitoring

on circumstances. It may be impossible to take any remedial action if the programme is such that retakes are impossible. On the other hand, if the trouble is overload distortion then this should be detected before it becomes very obvious and input levels can be reduced.

15.5 Tape speeds

There are certain 'trade-offs' between tape speed (the speed of the tape as it passes the heads) and quality. The greater the tape speed the better the high frequency response (because the recorded wavelengths are greater in relation to the head gaps). On the other hand the faster the tape the greater the cost, simply because more tape is used to record a programme. (In the professional world it is not usually acceptable to erase a tape and re-use it, and certainly not if there are edits held together by a special sticky tape. While this adhesive tape is very reliable there is a small chance that it might come apart when making another recording. The possible cost of starting a whole programme again, assuming this can be done, is much greater than the cost of a new tape.)

The standard professional tape reel is known as the *NAB spool* (NAB = National Association of Broadcasters) and contains 2400 feet (732 m) of tape on a '10-inch' (25.4 cm) spool. (Curiously the analogue tape world seems to prefer Imperial units: the standard speed is usually referred to as 15 inch/s rather than 38 cm/s and the tape itself is most commonly called 'quarter-inch tape', when it is 6.3 mm wide!)

A 10-inch NAB spool full of tape will run at 38 cm/s for 32 minutes. This means that a 30 minute programme can be recorded on it with a couple of minutes spare for test tones. The table below gives some of the more important specification features of a professional analogue tape machine.

Tape speeds	38, 19 (and possibly) 9.5 cm/s ±0.2%
Time to reach set speed	< 0.5 s (< means 'less than')
Rewind time	120 s for 2400 ft reel
Frequency response:	
at 38 cm/s	30 Hz to 18 kHz ±2 dB
	60 Hz to 15 kHz ±1 dB
at 19 cm/s	30 Hz to 15 kHz ±2 dB
	60 Hz to 12 kHz ±1 dB
Overall signal-to-noise ratio	61 dB at 38 cm/s
Bias and erase frequency	150 kHz

15.6 Tape editing

There are broadly two methods of editing analogue tape: *dub editing* and *cut and splice editing*. We'll take these in turn.

In dub editing ('dub' means copying, in this context) two or maybe three tape machines are needed. If, for example, a section of a recording is to be removed then everything *except the unwanted section* is copied to a second machine. Figure 15.5 shows a very simple application which could probably be best used for *compilation* work – putting different items together on one tape.

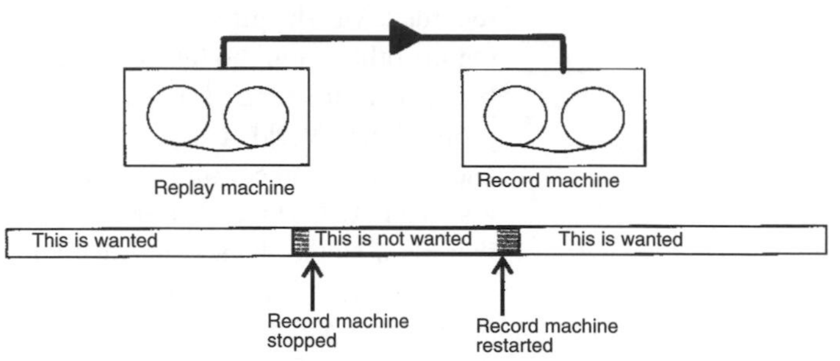

Figure 15.5 A simple editing system using two machines

A more useful application of dub editing is shown in Figure 15.6 where sections of two tapes are to be combined into one. For example, tracks 1, 3, 4 and 6 on tape 1 are to interleaved with tracks 2, 5, 7 and 8 on tape 2. The final version on machine 3 will have tracks 1 to 8 recorded on it.

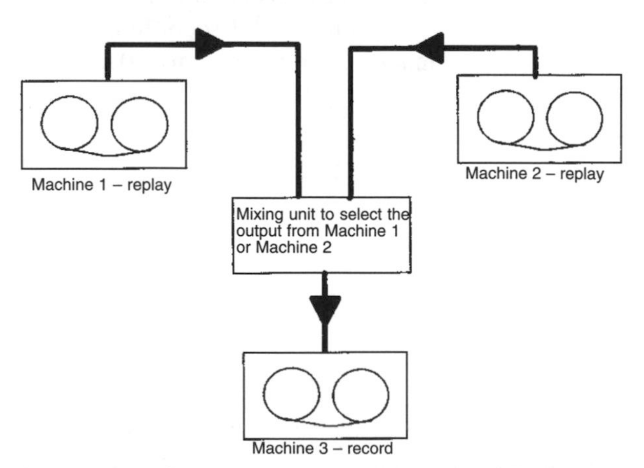

Figure 15.6 Outline of a more useful dub editing system

We shall compare the relative advantages and disadvantages of dub editing when we have looked at the cut and splice method.

In cut and splice editing the tape is physically cut and then joined up where needed. Three items of equipment are needed:

1. An editing block. This is made of metal and a section through it is shown on the left in Figure 15.7. The slot is some 10–12 cm long and the tape is held in it snugly and firmly but in such a way as not to damage it. An editing block is a piece of precision engineering and its cost is likely to reflect this! Slots in the block allow a razor blade to cut the tape at the correct angle.
2. A special single-edged razor blade. This needs to be sharp otherwise a ragged cut results. (*Used blades must be disposed of carefully. Special slotted plastic boxes are often available. Blades must* never *be allowed to lie around where they could present a risk to anyone.*)
3. Jointing tape. This is thin, flexible and adhesive on one side. Figure 15.7 (right) shows a piece of jointing tape correctly positioned over a joint in the tape. The mechanics of cut and splice editing are fairly quickly learned. The skilled part of any editing operation is knowing exactly where to make the edit!

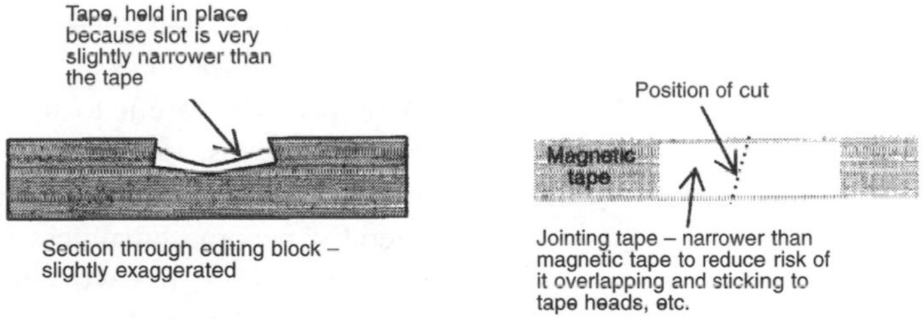

Figure 15.7 Section through an editing block (left). Position of jointing tape (right)

On the next page we give a comparison of the pros and cons of the two editing methods.

Finally, in this section, a few points about precise editing:

1. The exact point to edit can be found by 'rocking' the tape slowly to and fro past the replay head and listening to the output on a good loudspeaker. When this has been found the tape can be marked with a suitable wax pencil. With wax pencils great care has to be taken not to put any wax on to the heads.

Comparison of dub editing with cut and splice editing

Dub editing – advantages:

1. Much easier if considerable lengths of tape are involved, as for example, in a compilation tape made up of extracts each a few minutes duration.
2. The tape is not mutilated and can, in principle, be re-used.

Dub editing – disadvantages:

1. Costly in the sense that two or maybe three machines are tied up.
2. Precision editing may be impossible (although if an electronic timing track called time code is used precision is possible. We shall mention time code later.)

Cut and splice editing – advantages:

1. Only one tape machine is needed.
2. A skilled editor can handle edits of duration much less than a second.

Cut and splice editing – disadvantages:

1. The tape is weakened at each splice
2. Very cumbersome if sections of tape more than a very few metres in length are involved.

2. The start of a sound is easier to detect than the finish, so this is where the tape should be marked.
3. With speech be careful to allow reasonable gaps between edited sections, otherwise it sounds as if the speaker never stops to take a breath, and the listener can, quite literally, become breathless too!
4. Some speakers, not working from a script, may put in 'ums' and 'ers' from time to time. There may be a temptation to edit these out but there is a risk that in doing so an important characteristic of that person may be removed. This can sound unnatural if that person is well-known and the 'um' and 'er' tendency is associated with him or her. If the mannerisms are not too annoying it may be best to leave at least some of them in.
5. Music editing is difficult. One thing to watch is maintaining the rhythm (or beat) of the music. Any edit must preserve this.

15.7 Line-up

In an earlier chapter we mentioned the need for all equipment in a chain to be working to the same standards of signal level. This, of course, is applicable to tape machines as to much else. Different organizations tend to have slightly different procedures but we give here a typical one suitable for routine checks.

Step	*Action*
1	Check that meters (PPMs, etc.) are reading correctly. It is likely that a straightforward operation of a switch will feed line-up tone to the meter(s)
2	A tape for replay will probably have at least half a minute of line-up tone recorded on it. On replay this should indicate zero level on the meter. Alternatively replay a standard line-up tape. This may have a range of frequencies on it but for routine line-up the most important is several minutes of 1 kHz tone recorded at such a level that a correctly lined-up machine will replay the tone at exactly zero level (PPM '4'). If the meter does not indicate zero level output then adjustment of the replay gain control is needed. (This may be a job for a supervisor.) *Line-up tapes are very expensive and must always be treated carefully. Never record anything on them.* If the machine is to be used only for replay there is no need to go beyond Step 2. If it is to be used for recording then continue.
3	Lace the machine with the tape to be used for the recording; set the machine to record, feed zero-level tone to the input and check that the output is also at zero level. It may be necessary to check and adjust the bias but this should be left to a maintenance engineer or a supervisor.

15.8 At the end of a recording session

There is, inevitably, some paperwork to be done and when we've said what this is the reader is bound to see that it's important. Before that, though, there is the tape itself.

We implied a little earlier in this chapter that a tape for replay may have a section of line-up tone recorded on it at the start. This is important as replay may take place from an entirely different studio or location and the existence of this tone is vital for ensuring that the replay is at the correct level. Therefore before the recording starts there should be half a minute of this

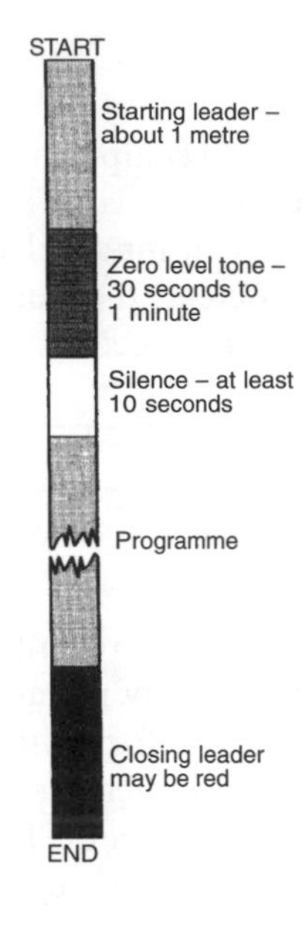

tone, followed by at least 10 seconds of silence before the recording begins. The reason for the silence is to prevent the effects of any *print-through* being heard. Print-through is the name given to the effect of programme material, especially if it is at a high level and the tape has been stored for an appreciable time, causing a corresponding magnetization in adjacent layers of tape. (The effect can often be heard as a *pre-echo* on cassettes.)

Then, after the recording, there is the process often called 'topping and tailing' – attaching suitable *leaders* of coloured, non-magnetizable tape to the start and end of the recording. (They're still called 'leaders' at the end of the tape!) Different organizations may use different colours but an example would be to have yellow leader – about a metre in length – at the start and red, again about a metre, at the finish. There should be no spaces between the starting leader and the tone nor between the end of the recording and the end leader. The diagram at the side illustrates all this, although some organizations may have additional tones at various frequencies in addition to the zero level tone we have shown.

Now the paperwork. *Every* recording should have with it complete documentation ('*recording report*') on, usually, a standard form that is kept in the box with the tape. This should include:

1. The title of the recording. (Obvious!)
2. The performers.
3. The duration of the recording, or each 'take' to within a second if possible. (This is the duration of the recorded material and shouldn't include any additional test tones.)
4. The tape speed.
5. Where the recording was made – for example the studio number or recording suite number.
6. The date when the recording was made.
7. The name of the recording engineer/operator.
8. Any special comments which may be helpful at the replay. 'Intentionally low level at start', 'slight microphone rumble 21 minutes in', are possibilities.
9. *In cues* and *out cues*. How the recording starts and finishes. If it is only music then the statements 'Starts: music, Ends: music' are about the only comments. However if there is speech then the opening words can be a helpful confirmation that the right tape is being

replayed. Similarly the closing words tell the replay operator to stand by to fade out the machine.

10. A tape may contain several retakes and it's important to indicate the extent to which each one is satisfactory, or not. Abbreviations such as 'FS' (false start), 'OK' (a useable take), 'NG' (no good) and a tick to indicate the chosen take are common.
11. Most organizations will have their own cataloging system and this must be adhered to. A fictitious but plausible example is:

    T15/03/95/43

    T15 = Tape, 15 inch/s; 03/95 being the month (March) and the year (1995), with '43' being the 43rd tape to be made that month. This number, whatever it is, should probably be written on a sticky label actually on the tape spool, also appear on the recording report, *and* go on an adhesive label on the box. Thus, if the tape, the box and the report all get separated it should be possible to identify each and bring them together again.

All the information in the report could be of the greatest importance at replay and it should never be neglected or skimped.

16

More about analogue recording

16.1 Cassettes

When compact cassettes appeared in the 1960s they were not taken seriously by the professional world because of their very low grade quality. This was largely due to the very narrow tape tracks resulting in poor signal-to-noise ratios. Also the early cassette machines were often inferior in performance. Since then the tape and the machines have improved out of all recognition and while cassettes do not give the same quality as a full size (*open reel*) machine, and still less than digital recording, nevertheless their performance can be perfectly adequate for some professional purposes, such as speech. (But even so, very compact digital recorders are replacing cassette machines.)

The general principles of cassettes and their machines are exactly the same as those of open reel machines and we shall not repeat them. It will be enough to give the important data about cassettes, starting with the track layout. Figure 16.1 shows how the tracks are arranged for cassette tape, while Figure 16.2 shows the layout for 'quarter-inch' tape.

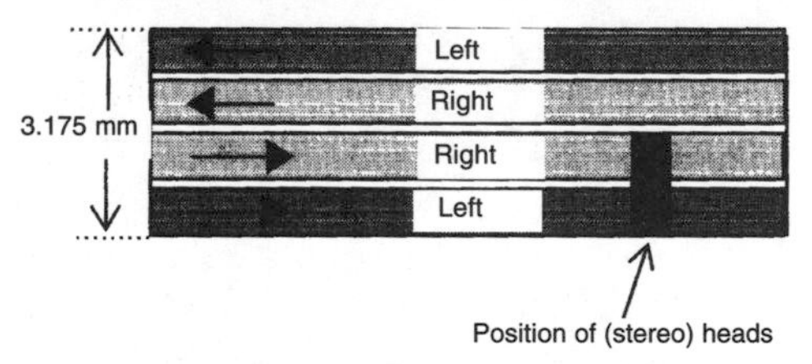

Figure 16.1 Cassette track layout

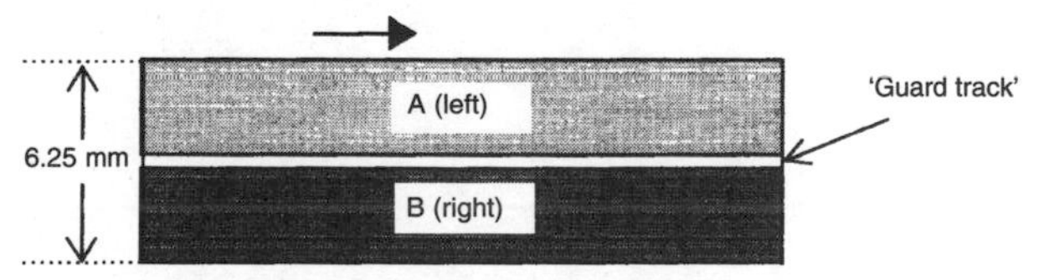

Figure 16.2 'Quarter-inch' tape tracks

One thing to notice is that there are four cassette tracks in half the width of conventional tape with its two stereo tracks. This helps to explain one reason for the relatively inferior quality of cassettes – each track is only a quarter of the width. As a result there are only 25% of the magnetic particles available for recording and replay compared with conventional tape. Or, put another way, the effects of tape hiss are going to be worse by a factor of four, in voltage terms. Translated into decibels this is equivalent to about 12 dB poorer signal-to-noise ratio. Added to this is the fact that cassette tape speed is low – 4.75 cm/s (1⅞ inch/s) – one eighth of the speed of open-reel tape, so it's not surprising that cassette quality is inferior! However, as we shall see later in this chapter, quite a lot can be done with *noise reduction systems* to improve the situation considerably.

A final point about cassette machines is that editing can only be by a dubbing process. Cut and splice devices aren't a serious possibility. To begin with, any cut will be across both sides of the tape and is most unlikely to be in the right place when the cassette is turned over! And also there's a serious risk of the splicing tape getting jammed in the mechanism.

A principal use for cassette machines in studios is for making copies of recordings for clients and performers to take away with them after a session.

16.2 Guard tracks and twin tracks

First, *guard tracks*. These are gaps which are intended to prevent any flux from one track being picked up by the head for another track – a condition known as *crosstalk*.

This is perhaps an appropriate point to distinguish between 'quarter-inch' stereo and *twin track* recordings. They might appear to be the same but in fact the professional world makes a definite distinction. Stereo we know about. One feature is that there is generally not a very fundamental difference between the A and B signals. They are, after all, parts of the same programme material. 'Twin track', on the other hand, means that entirely different material may be recorded on the two tracks. For example, a vocalist in a television recording might be singing to a backing which has been previously recorded on one track.

The physical difference between stereo and twin track machines is that the guard track is wider for the latter. A little crosstalk in stereo may not matter. In fact all it will do is to reduce slightly the width of the stereo image. Crosstalk with two entirely separate tracks is, of course, much more serious. Twin track machines can be easily distinguished because they have two faders, one for each track. A stereo machine should never be used for true twin-track work!

16.3 Cartridge machines

'Carts', as they are often called, are a kind of specialized cassette machine. They use a cassette containing an endless loop of 'quarter-inch' tape, running at 19 cm/s, so the quality is reasonably good. Their purpose is to provide a source of jingles, standard announcements or other routine sounds of relatively short duration.

The clever part is that special tones can be recorded on the tape in such a way that they are not reproduced in the output but provide automatic cueing. This means that when a jingle, say, has been played the tape continues and stops just before the *start* of the jingle. So, the next time the machine is operated the jingle is instantly available. To make sure that getting to the start of the item is as quick as possible cartridges are available in a wide variety of lengths – from as little as 10 seconds of tape up to about 10 minutes.

Analogue cartridge machines are being rapidly replaced by their digital equivalents, which we shall look at in another chapter.

16.4 Multi-track machines

The term is self-explanatory, at least in the analogue world. The aim is to be able to record or replay simultaneously a large number of entirely different tracks, each one being, to take an example, the output of a microphone or maybe a group of microphones covering an instrument or group of instruments in a band. Having made the recording the balancing of the tracks (the *mix-down*) can be done at a later stage. This technique is very widely used in the recording industry and to a considerable extent in broadcasting. The basic advantages are that the musicians do not need to be kept in the studio while the possibly time-consuming balance is carried out and also that complex processing of the tracks can be carried out without the pressures of a normal studio session. There is also a further advantage in that a singer, for example, may not be available on the day that the band is in the studio, so the band tracks can be laid on one day and the vocal track can be added later.

Multi-track machines may have up to about 32 tracks – 16 or 24 being quite common and the tape width is related to the number of tracks – '2 inch' (5 cm) is standard for analogue machines. It is usual, though, to reserve one track on analogue tape for *time code*, which is a digital signal carrying timing information. (We shall deal with time code more fully later.) Also, since time-code signals tend to be intrusive and crosstalk on to another track is not unknown, it is quite common practice to put the time code on an outside track, with the track next to it left blank as a kind of guard track. The point about having time code is that it makes it possible for a computer to control many of the operations, such as searching for a particular point on the tape.

Analogue multi-track machines almost invariably have a noise reduction system incorporated as the individual tracks are narrow and therefore don't have an inherently good signal-to-noise ratio.

Important features of any multi-track machine are the *sync outputs*. Suppose we have a vocalist singing to pre-recorded backing tracks. The vocalist will need to hear these, obviously, but if they are taken from the normal replay heads they will be delayed by a fraction of a second after the vocal track. The answer is to switch the record heads on these tracks to be replay heads and this removes any time delay between the replayed tracks and the vocal track. Such outputs are called, for very good reasons, 'sync outputs'. The quality of a sync output may not be quite as good because the record heads it's taken from are designed for recording and not for replay. However, this may not matter.

Digital multi-track machines are slowly replacing analogue ones but the process is likely to be fairly slow – the costs of both types are high and a change from one to the other represents a large capital outlay.

16.5 Noise reduction systems

There are many systems in existence and we shall look only at the ones the reader is most likely to meet. One of the best-known sets of noise reduction equipment is referred to as 'Dolby', after the inventor and originator. (The name Dolby is a registered trade mark.)

The first Dolby system was known as 'Dolby A' and was an

ingenious attempt to reduce the effects of noise entering a programme chain, either on a long cable link, or on a tape recording. With high quality digital satellite communications there is these days little need for a noise reduction system in long distance links, but Dolby A is still used on analogue multi-track tape machines, where, as we have seen, overall signal-to-noise ratios are apt to be poor because of the narrow tracks.

The basically very poor signal-to-noise ratio of cassettes has been greatly improved with the incorporation of Dolby B – a much simplified and consequently less expensive but nevertheless effective version of 'A'. There are others which we shall mention later in this chapter.

To begin with, all noise reduction devices are basically *companders*: they compress the dynamic range before the recording and then expand it afterwards. However there's more to it than just that. The dynamic range reduction needs to be done in such a way that the noise, typically tape hiss, which gets into the signal is not allowed to interfere permanently. Figure 16.3 should help to explain this. A compressor, the first part of the compander, reduces the dynamic range and the expander part, in pushing down the lower levels of the signal also pushes down the noise which has crept in during the recording process.

There are practical difficulties, in particular ensuring that both compressor and expander operate in precisely complementary ways: the amount of expansion must be exactly the same as the amount of compression.

The Dolby systems have an additional feature, namely that the companders only do their work when the signal levels justify it. For example if signal levels are high then it is likely that the

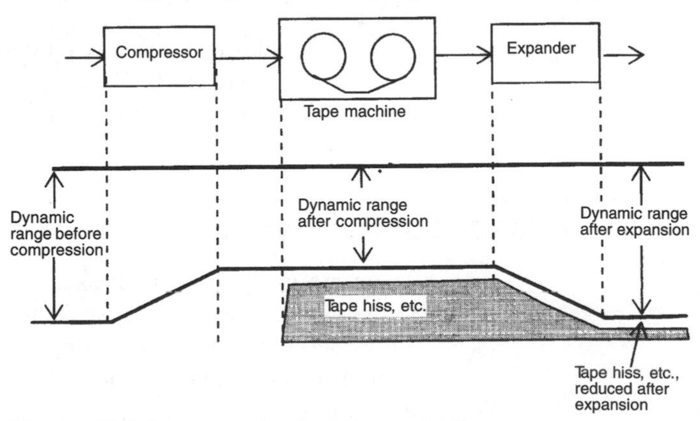

Figure 16.3 A compander in noise reduction

tape noise will be masked. In such circumstances no companding occurs. However, when signal levels are low, and the noise would not be masked, then companding takes place. In the complex Dolby A process there is yet another refinement in that the audio signal is divided up into four frequency bands, each with a compander operating independently of the others. This, of course, makes the Dolby A system expensive, especially when one remembers that a 32 track tape machine will need 32 sets of Dolby! A further point about the Dolby A system is that it is very dependent on accurate levels – a fairly obvious statement when one remembers that the circuitry detects the signal levels and only operates when the levels are low. Dolby A units produce their own special line-up tone – it has a distinctive warble sound – and this needs to be recorded at the start of all tapes which are Dolby A 'encoded'. Overall about 10 to 15 dB of noise reduction occurs. This may not seem a lot but one should bear in mind that it is used with the better quality analogue recording systems to start with.

The other noise reduction systems are, for the most part, versions of Dolby A, some simpler and others more complex. We will give a brief summary of each.

1. **Dolby B:** This was intended to reduce the effects of tape hiss on cassettes and therefore operates only on the higher frequencies. It can be thought of as a kind of top-lift circuit before recording with corresponding top-cut on replay. The frequency range over which it works is variable according to the amount of high frequency present. There is about 10 dB of tape hiss reduction.
2. **Dolby C:** This is also for use with cassettes. It's similar to Dolby B but is more complicated in that it divides the audio signal into two frequency bands and can give about 20 dB of noise reduction. However it is more level dependent than the B system.
3. **Dolby SR** ('spectral recording'): This is a complex system which has an action that depends upon three levels. At low levels there is a high recording level with a fixed equalization. At medium levels a range of variable filters is brought in which operate according to the frequencies present in the audio signal. At high signal levels gain reduction occurs but only in the frequency region of the high level signal. Up to about 24 dB of noise reduction can be achieved.
4. **dbx:** This is a fairly straightforward compander system. It

is interesting in that it uses an inaudible control signal to make sure that the compressor and expander are always matched. Noise reduction of 25 to 30 dB is claimed but it is possible for there to be unwanted audible effects as a result of the compander action.

16.6 Vinyl discs

We are told that King Charles II 'took an unconscionable time dying'. The same sort of remark could be made about vinyl discs! Not long after the use of compact discs (CDs) became widespread the end of the black gramophone record was confidently predicted but these discs are still around and there are even devotees who maintain that they have a better sound (whatever that means) than CDs. A mention in this book is justified on the grounds that vast numbers of important recordings exist on these discs, and while much archive material is being transferred to more durable media, there is still a place in broadcasting organizations for the traditional turntable.

We can assume that the reader is familiar with the basic principles of gramophone discs. All we will do here is to mention some of the more important features.

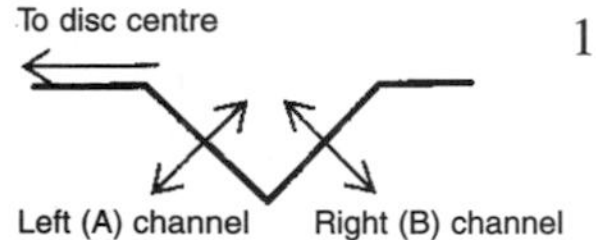

1. The groove of a stereo disc carries the A (left) channel information as a 45° movement in the left wall of the groove, as seen with the disc centre to the left. The B (right) information is carried similarly in the right groove – see the diagram in the margin.
2. The signal-to-noise ratio of a new disc is very good, 90 dB or more and comparable with a CD.
3. There is a deterioration with every playing. This is small and isn't likely to be noticed between successive playings, but it occurs nevertheless.
4. Discs should never be picked up or held with fingers on the grooves. Either they should be held at the rims, using two hands, or the other correct way, and it doesn't require large hands, is to support the disc with the fingers slightly spread out near the centre, on the label, and with the edge of the disc tucked into the soft flesh on the thumb at the first joint. In this way the disc is quite securely held and the grooves aren't touched.
5. Cleanliness is vitally important. Not only are the grooves

easily damaged, causing a distorted sound output, but dust which gets into the grooves can get embedded into the disc material by the pressure of the stylus. This can cause a background crackly hiss or, at worst, very audible clicks. A little top cut can reduce slightly the effects of dirt in the grooves but it is likely to be at the expense of the quality of the recorded sound.

Vinyl discs are often used because they are important recordings and therefore have a rarity value. It's worth taking extra care of them.

17

An introduction to digital audio

17.1 What's the point?

We've seen that analogue systems such as magnetic tape and vinyl discs are somewhat prone to uncorrectable errors such as drop-outs, tape hiss, scratches and so on. Also transmission systems – where the audio signal is sent by radio waves or wires – can suffer from various forms of interference. The question is, can the signals be converted into another form which is not so seriously affected by system errors?

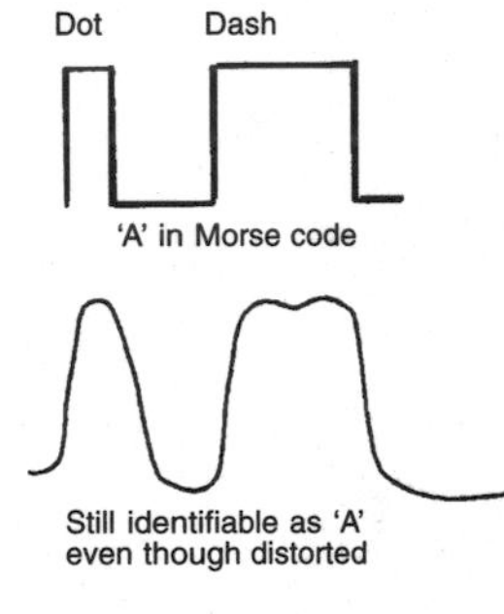

The answer is, of course, Yes. (Otherwise we wouldn't have asked the question!) It's interesting to look briefly at Morse code. With its use of dots and dashes it is possible to decode a message correctly even if a considerable amount of impairment of the signal has taken place.

A code with some similarities to Morse code is used as the basis for digital audio (and also, incidentally, for digital television) and we will next look at the steps which are taken to convert an audio signal into a reasonably incorruptible code. There are three fundamental stages: we shall call these *sampling*, *quantizing* and *coding* and they are performed by a device, in the form of a chip, called an *analogue-to-digital converter* or ADC.

The conversion back to analogue is by a *digital-to-analogue converter* or DAC.

17.2 Sampling and quantizing

The first step in the ADC is to examine minutely each little piece of the audio waveform – to take *samples* of it. Figure 17.1 illustrates this.

In the quantizing stage, which we'll come to shortly, we shall see that the 'height' of each sample is measured and this measurement will, at a much later stage – *decoding* – be used to reconstitute the audio waveform. It should be obvious, then,

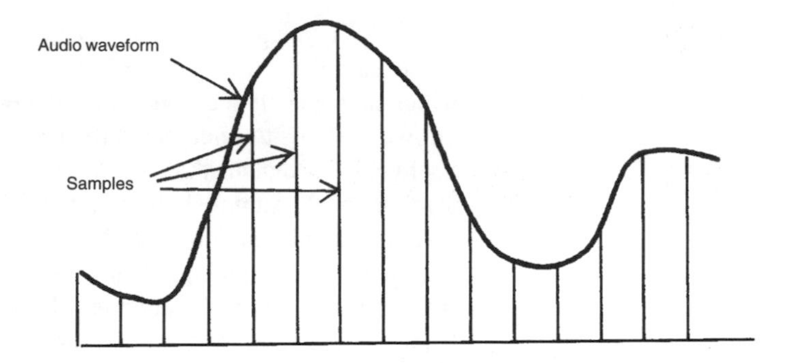

Figure 17.1 Sampling

that the samples must be very close together. Or, to put it another way, the *sampling rate* must be high. A good rule is that the sampling rate must be about 2.2 times the highest audio frequency, and if our highest audio frequency is to be about 18 kHz, then the sampling rate must be 2.2 x 18 000 which is almost 40 kHz. (In the UK the BBC's distribution of radio signals to transmitters uses a sampling rate of 32 kHz, but this still gives an upper audio frequency of nearly 16 kHz.)

There are two standard sampling rates used in the professional world, apart from the 32 kHz we have just mentioned:

1. 44.1 kHz, which is used for most digital audio such as CDs.
2. 48 kHz, which was intended to be the true professional standard with 44.1 kHz for the domestic world of 'consumer' products. However so much professional equipment today uses 44.1 kHz that it can be regarded as the norm. Having carried out the sampling process we can now look at *quantizing*, which is shown in Figure 17.2.

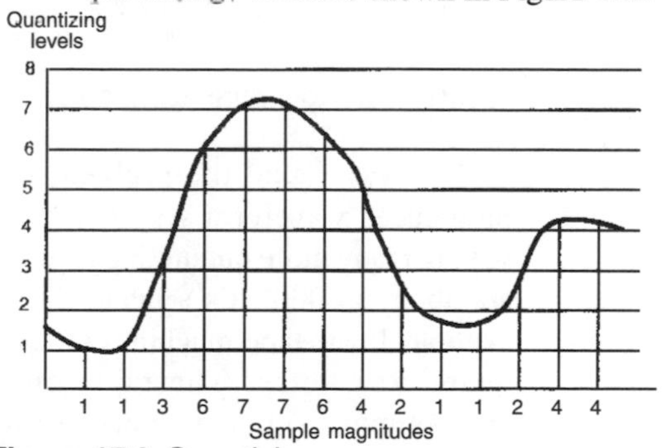

Figure 17.2 Quantizing

It helps to imagine a ruler being used to measure the amplitude of the samples. In Figure 17.2 the ruler has eight graduations which as we shall see is nothing like enough for good quality, but will help us to understand the process. The imaginary ruler is slightly special in that we can only measure to the nearest level below the sample height. In the diagram we use this law to give the sample magnitudes. (If we were to say that the eighth sample was 4.5 units high then we'd really be using a ruler with far more than eight graduations. The word 'quantizing' comes from physics and means 'a definite and fixed quantity'.)

Now let's look at the decoding process where the numbers, which in this example are 1, 1, 3, 6, 7, 7, 6, 4, 2, 1, 1, 2, 2, 4 and 4, are used to establish the height of the recovered waveform. Figure 17.3 shows what happens. We've superimposed the original audio signal and it's clear that what we have decoded bears only a vague similarity to the original. The gaps between the original and the decoded signal are perceived as a noise, known not surprisingly as *quantizing noise*. It's often said to have a 'granular' quality, which is a very fair description. It sounds 'gritty'.

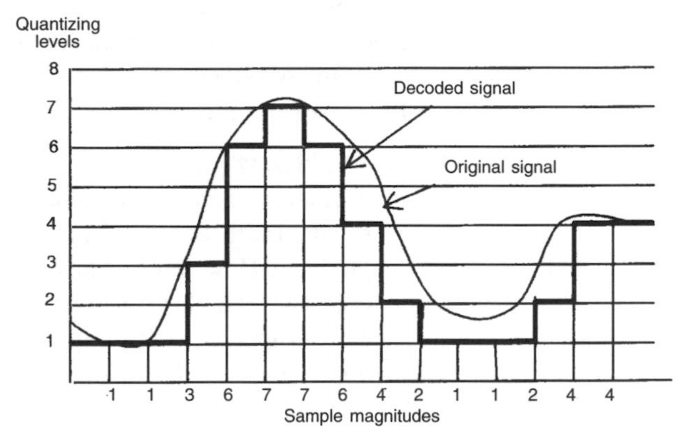

Figure 17.3 How quantizing noise arises

The way round the problem is to make the steps in the reproduced waveform very much smaller, and this is done by having many more quantizing levels. In a CD, for instance, there are about 65 000! It's a matter of signal-to-noise ratios again. With analogue tape machines the noise was tape hiss. Now it's quantizing noise. The table on the next page gives the approximate signal-to-noise ratios for various values of quantizing levels. The reader is invited to look for an interesting relationship! (*Hint*: Compare the increases in the two columns.)

Number of quantizing levels	*Signal-to-noise ratio (dB)*
64	25
128	31
256	37
512	43
1 024	49
2 048	55
4 096	61
8 192	67
16 384	73
32 768	79
65 536	85

So, what was interesting? Well, in the first column each number was doubled in the line below, while in the second column each number increased by 6. The reason is that if we double the number of quantizing levels the spacing between each is halved; thus the quantizing noise amplitude is also halved which means that it is reduced by 6 dB.

A second observation is that we have to go to a very large number of quantizing levels to get a good signal-to-noise ratio. In the table we've gone to 65 536, but even that is not enough for some purposes.

A third observation might be that we're having to deal with very large numbers. Sampling is going to take place at about 44 000 times a second, and each sample in a good quality system is going be a number which can be as great as 65 536.

Fortunately there is a way of reducing these numbers – *binary arithmetic* – which is the third job of the ADC.

17.3 Binary arithmetic

The reader may well have come across this at school, but if not, or for those who have forgotten it, here is an outline.

It's true to say, perhaps surprisingly, that it is the easiest possible type of arithmetic. If that doesn't seem to be the case then it's only because we're not used to it! Our usual scale of arithmetic – *decimal arithmetic* – is on a scale of 10 because we have 10 fingers (counting thumbs as fingers for the time being!)

to count with. We start at 0 and go up to 9, then we put a 1 on the left and go from 0 to 9 again (10 to 19), then put a 2 on the left (20 to 29), and so on.

Now suppose we had only our fists to count with. We'd go 0, then 1, and then we'd have to put a 1 on the left and go from 0 to 1 again, and so on. The only digits we need are 0 and 1.

Decimal number	*Binary number*
0	0
1	1
2	10
3	11
4	100
5	101
6	110
7	111
8	1000
•	•
•	•
•	•

The real importance of this, as far as we're concerned, is that if the very big numbers we're having to deal with can be converted into binary (and they can, quite easily, with modern electronics) then a 1 can be thought of as a voltage ON and a 0 as a voltage OFF. And if we represent these voltages by electrical pulses we have a code which is not too different from the Morse code idea we thought about at the start of the chapter. Figure 17.4 shows the kind of electrical waveform corresponding to 1s and 0s.

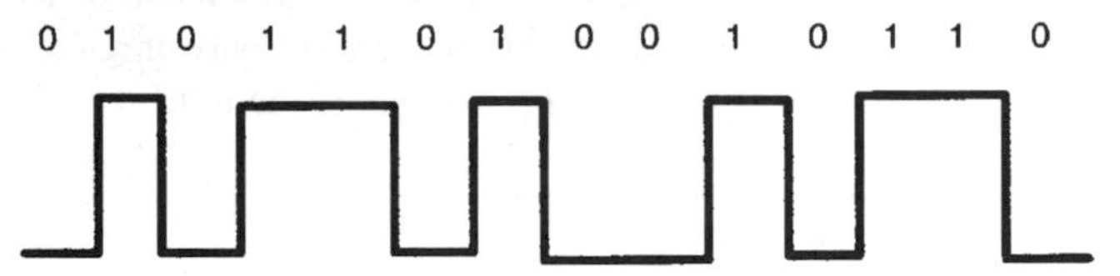

Figure 17.4 1s and 0s as a waveform

Terminology: The 1s and 0s are called ***binary digits*** or *bits*. Thus we speak of a '10-bit system', a '16-bit system' and so on. Figure 17.4 represents a 14-bit system.

In future, then, we shall refer to the number of bits rather than the number of quantizing levels. The next table is a repeat of the earlier one but with an extra column to show the number of bits.

Number of quantizing levels	*Number of bits*	*Signal-to-noise ratio (dB)*
64	6	25
128	7	31
256	8	37
512	9	43
1 024	10	49
2 048	11	55
4 096	12	61
8 192	13	67
16 384	14	73
32 768	15	79
65 536	16	85

A useful figure to remember is that 16 bits can give a very respectable signal-to-noise ratio. (The signal-to-noise ratios we quote here may differ from figures the reader might come across elsewhere – it depends on how the signal and the noise are measured. This is why, earlier, we used the word 'approximate' about signal-to-noise ratios.)

Before we go any further we should see one of the great advantages of digital signals. We've said that pulses can be severely distorted but can still be recognized, provided the distortion isn't excessive. But it's more than that. Provided an electronic device can recognize a pulse, even if it's far from perfect, then a new pulse can be created. Figure 17.5 shows this.

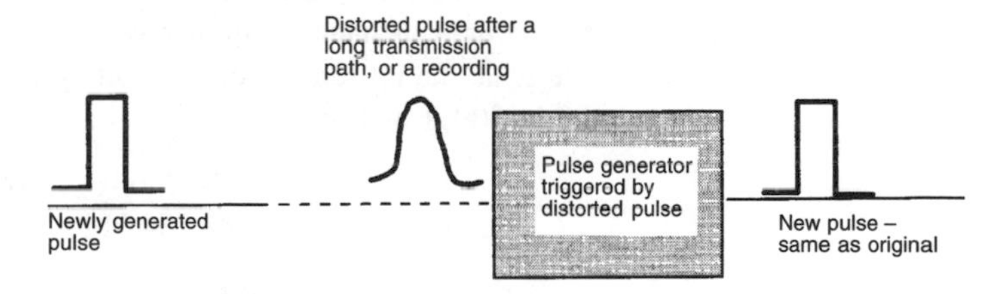

Figure 17.5 Regeneration of pulses

It might seem as if the process could go on indefinitely, but unfortunately this is not so. Sooner or later (hopefully later!) some pulses are going to be so distorted that they just cannot be recognized, and then distortion of the wanted signal becomes a possibility. Or, a 'spike' of interference may occur and be recognized, wrongly, as a pulse. Figure 17.6 illustrates these points.

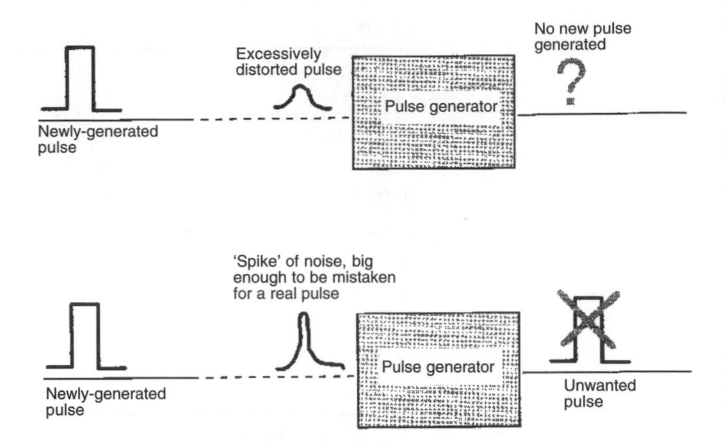

Figure 17.6 Pulse regeneration can be fallible!

We do, though, have another trick up our sleeve, which we shall look at shortly – *error correction* – in other words even when individual bits have been lost, or false ones created, it may still be possible to put things right – and this is something which is quite impossible in an analogue system.

17.4 Bit rate

This simply means the number of bits being produced, recorded, transmitted or handled in any way in a second. We can easily see what the bit rate is for a 16-bit system with a sampling rate of 44 000. It will be:

$$16 \times 44\,000 = 704\,000$$
or 704 kbit/s.

Perhaps we should think about this for a moment. Analogue tape will cope with frequencies up to about 18 or 20 kHz. How are we ever going to record frequencies up to around 700 kHz? And that's for mono! Stereo is going to need twice that, or around 1.4 MHz! This can be a major problem with digital audio, but, as ever, these difficulties are solvable and we shall see how digital recording handles these very high bit rates.

In the meantime we can mention that there are various methods of reducing the bit rate, some of them very complicated. Here we will do no more than give the briefest of outline. The interested reader can find out more from the book list in Chapter 23.

NICAM. There are two versions of this. One, properly called NICAM 728, is used for the transmission of television sound and will be outlined later. Here we are referring to the earlier NICAM used in the UK to send high quality audio signals around internal distribution systems to transmitters. The letters stand for Near Instantaneous Audio Companding Multiplex. The system starts with a 14-bit digital signal, which is quite good enough for material which is going to be transmitted and then picked up by domestic radios, and makes use of only the 10 most important bits. Figure 17.7 shows the idea.

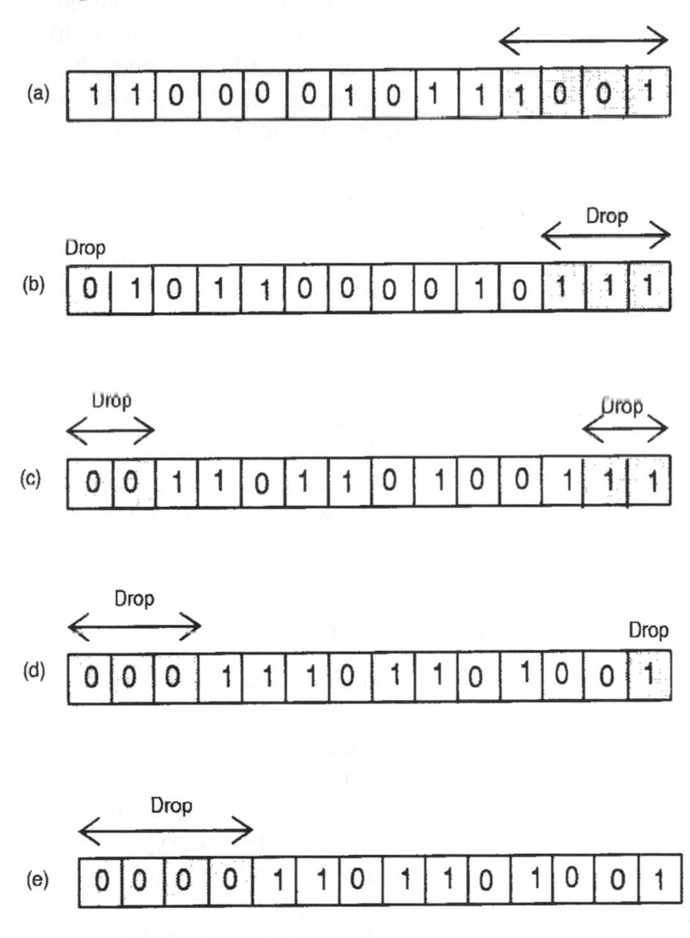

Figure 17.7 NICAM bit-reduction

If the sample is a high level one the last four digits can be dropped because they represent parts of the sample that are some 50 dB below the rest. Only the first 10 digits will be transmitted as in Figure 17.7(a). At the decoding stage the missing four will be replaced by four 0s. In (b) the amplitude of the sample is a little less so the first digit will be a 0. That can be dropped as can the last three digits, and so on. (e) represents a relatively small amplitude sample and the first four digits are zeros anyway, so dropping them and re-inserting zeros at decoding doesn't in fact degrade the signal at all.

Overall then, a 14-bit signal can be transmitted as a 10-bit one with almost no detectable reduction in audible quality. There does, of course, have to be a little extra information to tell the decoder which level is being handled and therefore how many 0s to insert where, but that can be done by sending two digits at intervals (once every 32 samples, in fact).

A further method of reducing the bit-rate is known as *PASC* (*precision adaptive subband coding*) or *perceptual coding*. This is radically different from NICAM and in effect mimics the behaviour of the human ear. We said in an early chapter that we cannot hear sounds below the threshold of audibility. Therefore, the argument goes, we don't need to process those 'sounds' at all. Also there is a rather complicated effect called *noise masking* which means that we cannot hear certain frequencies in the presence of others. Advanced circuitry makes use of these effects and it becomes possible to throw away up to 80% of the original digital data. It might be that direct comparison of the original signal with the PASC encoded one could show a very slight degradation in the latter but it will be small.

17.5 Error correction

Correction of analogue errors is either not very effective or it's very time-consuming. For example 'scratch' on an old 78 disc can be minimized, actually very effectively, by special computer techniques but this is slow work. Tape hiss can be made less objectionable with a little top-cut, and so on, but these are really methods of alleviating symptoms, like giving aspirin to someone with a broken leg. It helps at the time but it doesn't treat the broken leg!

In the digital domain it is fairly easy, first to detect errors, and secondly to go a long way towards compensating for them, maybe even correcting them completely. There are many types of error detection and correction but here we shall deal with just two as examples.

1. **Parity:** This is a simple but effective way of dealing with relatively minor errors and it is very widely used. At the encoding ('sending') end the 1s in the sample are counted and a further digit is added to make an even number of 1s. This is known as *even parity*. (Some systems use *odd parity*.) The chart below may help to make this clearer.

Original sample	*Number of 1s*	*Add parity bit?*	*New sample*
0101110010110101	9	Yes	0101110010110101 1
1011000101100110	8	No	1011000101100110 0
1110110110101010	10	No	1110110110101010 0
0010010100110101	7	Yes	0010010100110101 1

Every transmitted or recorded sample now has an even number of 1s. At the decoding stage the number of 1s is counted. If there is still an even number it is assumed that all is well. However an odd number of ones shows that something has gone wrong. This is error detection. A common technique when an error is detected in this way is to repeat the previous sample, which is, after all, only about 1/44 000 of a second away, so the effect is not very likely to be audible.

The fallacy with simple parity is that if two errors occur they will not be detected. However with a reliable system the chances of two errors in one sample is small.

2. **Scatter codes:** These give a high degree of correction for major losses in data, such as a drop-out in tape, or a flaw in the pressing of a CD. The principle is shown in Figure 17.8.

   The bits in the original sample are scattered according to a pre-arranged code. If there is a major error (a 'burst error') then on reassembling the bits the effects of the error are distributed and can be handled by some form of parity. (The more complex parity systems can even go so far as to reinsert the correct bits!) Scatter systems are so effective that on a CD for instance it is possible to have a hole in the disc up to about 2 mm diameter. The reader is strongly advised *not* to try any experiments, at least not with favourite discs!

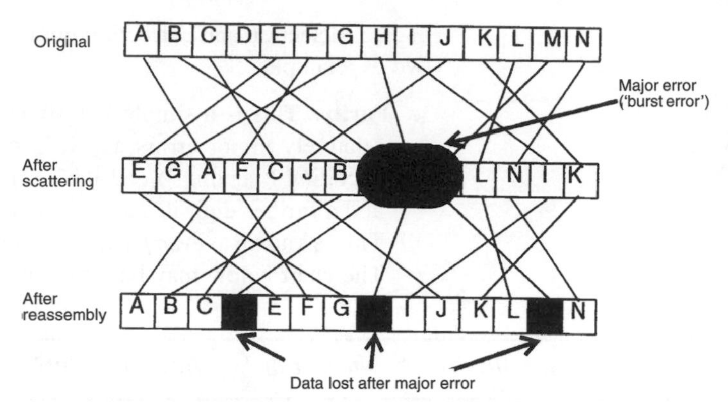

Figure 17.8 Effects of a burst error minimized

17.6 Drawbacks of digital audio

Are there any? Well, yes, but it has to be said that the advantages are enormous. We've seen that very high audio quality is achievable; errors can be detected and corrected; the costs these days are generally very reasonable – sometimes much cheaper than analogue equivalents; and it is possible to perform a great variety of processing 'tricks' such as artificial reverberation and pitch changing. We shall look at some of these in a later chapter.

The two principal drawbacks, which are not too serious provided one knows about them are:

1. If there is trouble it can be catastrophic! For example with analogue audio a major overload will result in severe and unpleasant distortion but the digital equivalent, assuming that the problem is too great to be handled by the error correction strategies we have mentioned, may be a really loud bang in the loudspeakers.
2. Direct monitoring of digital signals is impossible. The signals have to be converted to analogue form first. This is not too serious a difficulty but is worth mentioning.

18

Recording with digits

Analogue recording and replay using tape and disc are both relatively simple – after all, magnetic recording has been in existence for over 50 years and disc recording has been around for a century so techniques developed as long ago as that are not likely to be very complicated. Digital audio grew up with advanced electronics in the form of chips – digital audio is only possible because of advanced electronics – so it's not surprising that digital recording, whether on tape or disc, is a complex matter. The reader needn't worry! Here we shall look only at the broad principles. In any case it takes a complete book to go in detail into either digital tape or digital disc.

We shall deal first with the two main methods of recording onto magnetic tape and then look at compact discs (CDs) with a final glance at other and more recent developments.

18.1 DAT (digital audio tape)

The problem with recording or replaying using magnetic tape for digital audio is the *writing speed* – the speed with which the tape moves past the heads. The reader will remember that the maximum frequency that can be put on to the tape in an analogue recorder is around 18 to 20 kHz, using reasonable tape speeds. The difficulty is that above those sort of frequencies, and with normal tape speeds, the recorded wavelengths become very short indeed. It then becomes difficult to replay them and also the very short magnets tend to demagnetize themselves.

The bit rate with digital audio is around 700 kbit/s for mono and double that for stereo. Actually the bit rate isn't necessarily quite the same as the frequency but for most purposes the two can be regarded as being the same thing. In other words, instead of coping with frequencies up to around 20 kHz, with analogue, we've now got to deal with about 1.4 MHz with digital stereo –

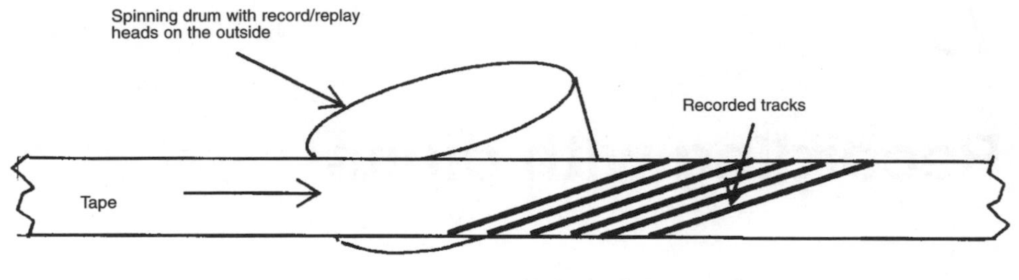

Figure 18.1 Slant head recording

roughly 70 times as much! A possible answer to the problem would be to increase the tape speed in proportion, but that would mean a full spool of analogue tape would have less than half a minute of recording time!

A vastly better answer had been found already in recording television on tape. Colour television signals go up to over 5 MHz and someone astutely realized that the necessary writing speed could be provided if both the heads and the tape moved. The head, or heads, are fixed into a drum which rotates quite fast and is on a slant to the tape. Figure 18.1 shows the idea, which is, by the way, the basis of the domestic video recorder.

With the head drum rotating fast the tape itself can move relatively slowly – a few centimetres per second. The tracks, if we could see them would appear slanted – the slope in Figure 18.1 is exaggerated for simplicity.

To ensure that the heads are in contact with the tape for long enough there is a partial 'wrap' round the head drum, as in Figure 18.2.

Of course, the process is much more complicated than this. For example there has to be a means of ensuring that on replay the head aligns exactly with the track it is intended to be covering, and the tracks are very narrow, about 14 μm, which is about one seventieth of a millimeter! And, with all this, the DAT cassettes

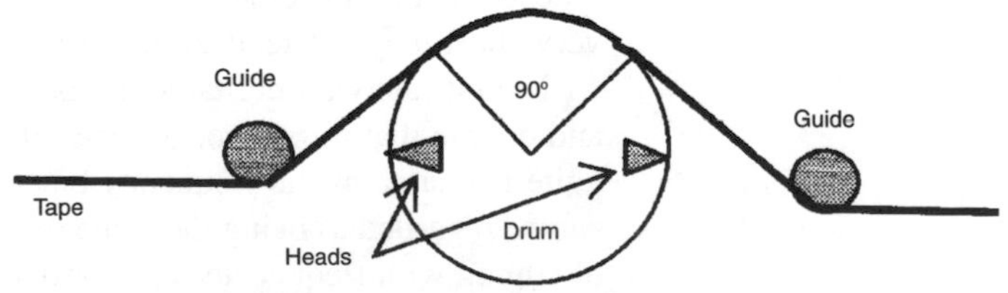

Figure 18.2 Typical head wrap in a DAT system

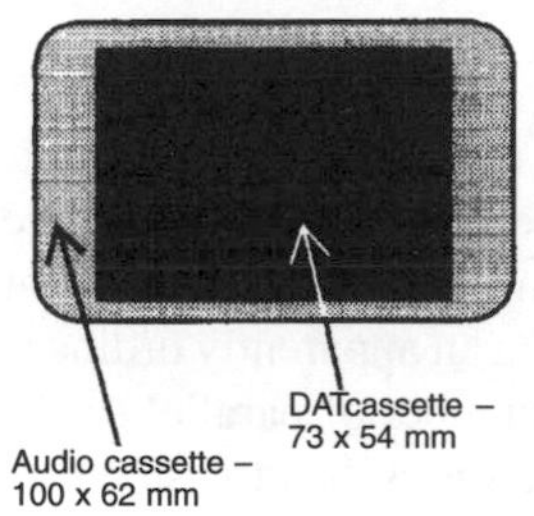

are smaller than an ordinary audio cassette and have a playing time of at least 2 hours, compared with the 45 minutes (60 minutes but with a risk of the tape stretching) on each side of an audio cassette.

The table below gives some of the more important facts about DAT with corresponding audio cassette details for comparison.

A look at the table suggests that DAT is almost too good to be true, especially when it is remembered that very full error detection and correction systems are incorporated. What, the reader may ask, are the drawbacks? Very few, in fact. With the first DAT machines there were problems with reliability and compatibility. The latter meant that it was possible to take a DAT tape recorded on one machine to find that it wouldn't play on another. However such problems have been solved with improved manufacturing tolerances.

	DAT	*Audio cassette*
Tape width	3.81 mm	3.81 mm
Tape speed	8.15 mm/s	42.5 mm/s
Effective head/tape speed	3.1 m/s	Equals tape speed
Angle of tracks	6° approx.	N/A
Recording time	*2 hours	Normally 45 min/side
Drum speed	*2000 r.p.m.	N/A
Frequency response	Up to 20 kHz	Up to 16 kHz, maybe a little more
Signal-to-noise ratio	90 dB	Around 60 dB with noise reduction

*These can be different. For example many DAT machines can be switched to 32 kHz sampling frequency instead of the normal 44.1 kHz. This can give 3 hours of recording, but of course at a slightly poorer quality.

The main difficulty is in editing. Rapid and accurate editing is possible but it needs extremely specialized and rather expensive equipment. Once the equipment is there, though, a computer screen can give a display showing the actual (analogue) waveforms of the audio signal and editing becomes quick and easy.

Before leaving DAT it might be worth mentioning that the first low-cost digital recorders were in fact domestic video recorders. They worked in conjunction with a unit which carried out the analogue-to-digital conversion (and the reverse on replay) and also processed the digital signal so that it was in the form of a video signal that the VCR would accept. To manage all this it was found to be convenient to adopt a sampling rate which was not exactly 44 kHz but 44.1 kHz – and that's how the '0.1' sneaked in!

18.2 Stationary head recording

The DAT system, as we've seen, gets round the very high bit-rate problem by increasing the effective tape-head speed. There is an alternative solution to the problem and that is to have what appears to be ordinary tape, travelling on an apparently ordinary machine at a normal speed, but has a number of parallel tracks of digital audio. There are several stationary head systems in existence. To the eye they look very much like analogue machines – the spools are similar, the tape transport mechanism is similar, and so on. The tape, though, has to be different. Even though the digital data are spread over several tracks this in itself is not enough to produce satisfactory digital recording. The recorded wavelengths are still very small and the tape needs to be thin and very flexible so that it maintains good contact with the heads.

One advantage of stationary head over DAT type machines is that razor blade editing is possible. This is where we find a very good use for the scatter codes we referred to earlier. A razor blade cut across the tape, and a subsequent splicing together of two lengths of the tape, results in massive losses of data, but these can be compensated for by scattering and parity methods.

There is an additional problem. To find accurately a point to cut the tape means 'rocking' the tape to and fro over the replay head. This is fine with analogue tape: the editor hears the programme material reduced in both speed and pitch, but can nevertheless identify the wanted edit point. Slow movement of a digital tape cannot result in any audio output, so the problem is solved by laying an analogue track along the tape. This can then be used to find the edit points.

18.3 Digital recording – advantages and disadvantages

Disadvantages

Very few, really. When the first digital recorders appeared they tended to be expensive, and this is still tending to be true of stationary head recorders. However prices have fallen, in common with most other electronic devices where silicon chips are used extensively, because once the chips have been designed their production in volume is cheap. Present day recorders, certainly if of the DAT type, are much cheaper than their analogue equivalents.

The main disadvantage, which applies only to DAT systems, is that editing requires expensive auxiliary equipment – but even the prices of these are much less than when they first appeared.

Advantages

1. Much better quality than analogue. The frequency response of a digital machine may not be vastly better than that of an analogue one, but the signal-to-noise ratio is a big improvement, some 90 dB compared with around 60 dB. True, noise reduction systems can make analogue performance a lot better, approaching digital standards, but these don't help greatly with the next advantage.
2. Repeated copying results in far less degradation than is the case with analogue. This is because digits which become impaired in a copying operation are 'cloned' for the next generation copy. Analogue tapes show detectable impairment after perhaps three or four generations. Digital tapes can be copied many more times – not an infinite number because eventually even the best error correction methods will fail to put right *all* the mistakes.

   It's just as well that copying has, up to a point, a minimal effect because digital editing involves a fair amount of copying.
3. Things like drop-outs can be corrected.
4. Machines don't need to be carefully lined-up like analogue machines. As long as a pulse can be identified as a digit then a new digit can be reconstituted. Bias adjustment is not necessary for instance.
5. Print-through, which can sometimes be a problem with analogue tape, doesn't matter with digits. This is not to say that print-through doesn't occur, but if it does the 'printed-through' digits will be of such low amplitude that they won't be recognized as digits.
6. Wow and flutter can be eliminated by the quite simple expedient of have a memory system which can act as a reservoir. No matter what rate (within limits!) the digits go into the store they can be extracted at the correct rate. Figure 18.3 shows the idea.

To summarize we could say that digital tape recording is electronically far more complex than analogue, but mechanically it is simpler. There still has to be precision mechanical

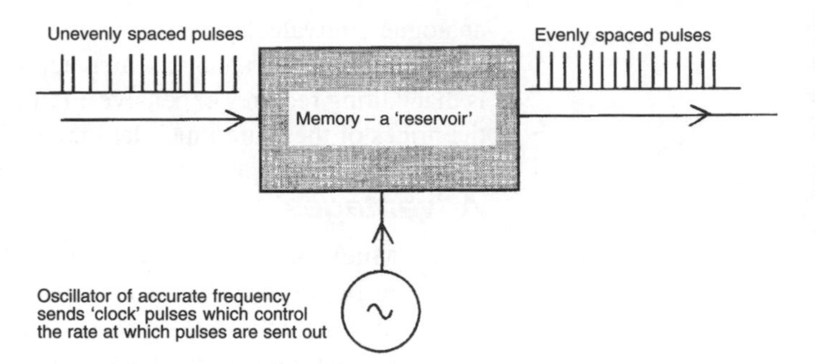

Figure 18.3 Removal of wow and flutter effects

engineering in producing, for example, the rotating head drum in a DAT machine, but there can be more tolerance in things like tape speed control, because if there are errors the electronics can be used to correct them.

For completeness this section ought to include a brief mention of the *digital compact cassette*, (DCC). This is a stationary head system and uses cassettes which are exactly the same size as the familiar audio cassette – indeed, conventional audio cassettes can be played on a DCC player. There are eight tracks of digital data and that fact, combined with the slow tape speed (4.76 cm/s) means that a great deal of data compression has to be used (PASC – see Chapter 17).

18.4 Compact discs (CD)

The CD is the device that has brought digital quality into almost everybody's home, and it's been made possible because of an optical device called a *laser*. (Laser is an acronym for light amplification by stimulated emission of radiation – which doesn't actually tell you much!) Basically it's the optical equivalent of a resonating pipe, which we explained in Chapter 6. And just as a pipe can produce a very pure sound, so an optical resonator can produce a very pure light, and by pure we mean just one wavelength.

The significance of this is that the digital data have to be very tightly packed and so the light beam needed to 'read' it must be very accurately focused. Ordinary so-called 'white' light, as we know, is made up of a number of different wavelengths. A lens bends different wavelengths by different amounts, an effect one can easily see with a simple magnifying glass, when there is marked spurious colouring, particularly near the edges of the image. Carefully designed and constructed compound lenses can reduce greatly this chromatic aberration so that for ordinary television and photographic purposes the spurious colouring is negligible. But such lenses are large and expensive and it's doubtful if they could ever be made small and light enough in weight to have the focusing abilities needed for CD recording and replaying.

The 16-bit data with a sampling rate of 44.1 kHz are etched on to a master disc using a laser. The laser beam is modulated ('switched' on and off) according to the digital signal. Then moulds are taken of this master and finally the production discs are made using a process known as *injection moulding* in which the hot plastic material is forced into the moulds. (This operation is often called *pressing* by analogy with the production of vinyl discs.) The surface with the digital data is made reflective and finally covered with protective glass. Figure 18.4 shows the minute size of the *pits* (the 1s and 0s). The surrounding flat part is called the *land*.

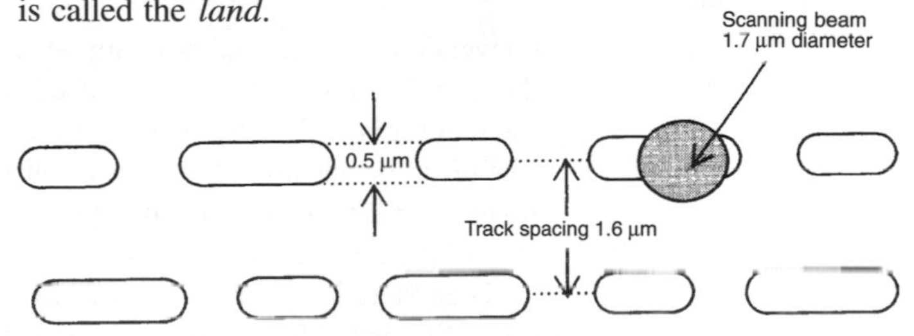

Figure 18.4 CD tracks

It's not easy to grasp the minuteness of the pits. It might help to visualize things if we suppose that a UK 5 pence piece represented a pit. Then on the same scale the complete CD would be 3.6 km across, or about two and a quarter miles!

Figure 18.5 shows, in slightly simplified form, the optics of a CD player. The laser is a low-power one – a high power laser could burn the tracks on the CD! Its beam is focused on to the track having passed through a half-silvered mirror.

This mirror allows half the light falling on it to pass through and half to be reflected. It means, of course, that half the light from the laser as it comes down to the mirror is reflected away to the left and lost. The other half goes on to the disc and returns, modulated by the pits. Half of that will pass through the mirror but the remainder goes to the right and is focused on to the sensor cell, the electrical output of which goes off for processing and

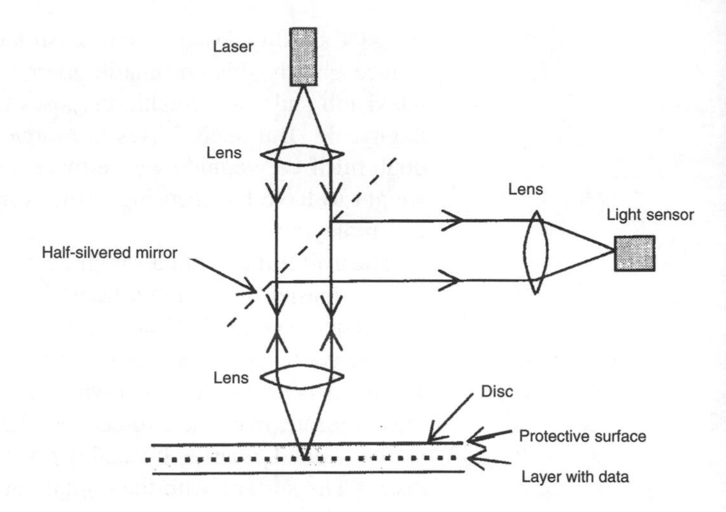

Figure 18.5 The optical system of a CD player

conversion into an analogue signal. It's interesting to see that the laser beam is accurately focused into a 1.7 μm spot on the track but it is still fairly wide at the disc surface – about 1 mm. The result is that small particles of dust on the surface are out of focus and therefore don't cause a significant amount of incorrect signal to enter the light sensor.

Then there is the problem of following the tracks. With vinyl discs it's easy – the stylus sits in a groove. With CDs there's no such groove to follow so instead an ingenious optical system does the tracking. Figure 18.6 shows the idea.

The laser beam is cleverly split into three by a device called a *diffraction grating* which, for the sake of clarity, has been omitted in Figure 18.5. We can call the three beams M for the

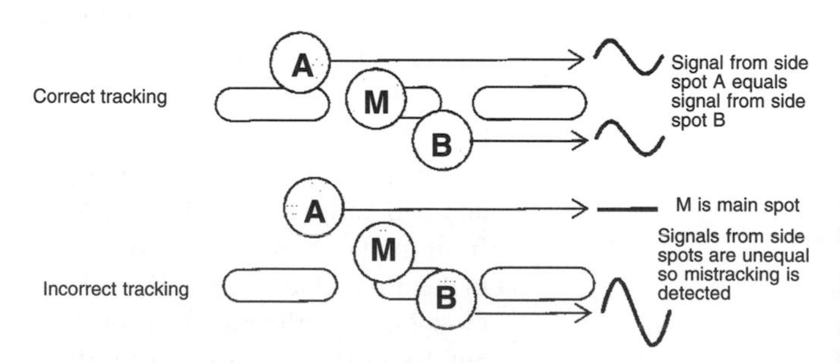

Figure 18.6 CD player optical tracking method

main beam which reads the track data and A and B for the side beams. If the main beam is correctly aligned with the track then A and B pick up small but equal amounts of the track. The A and B outputs are compared and, if they are equal, as they are in the upper diagram, then no action needs to be taken. If, though, the beam wanders slightly, so that, as in the lower diagram, the B output is much greater than the A output then the beam is steered back on to the track.

CDs differ from vinyl discs in other ways than those we have mentioned. To begin with the rate of rotation of a vinyl disc is constant – 78, 45 or 33.33 r.p.m. CDs have a variable rotation because it's important that that track data are read at a constant rate. Consequently the rotation is faster at the inside tracks than at the outside, so the velocity of the beam on the track is constant (normally 1.2 m/s).

Secondly the tracks are read from the inside to the outside – opposite to vinyl discs. The reason for this is that different diameter discs can be played without the need for complicated arrangements to detect the outside edge of the disc (as has had to be the case with automatic players of vinyl discs).

General facts about CDs	
Disc diameter	120 mm
Rotation speed	Varies according to track being read: about 500 r.p.m. at innermost tracks to about 200 r.p.m. at outside tracks
Track speed	1.2 m/s
Number of tracks (max.)	20 625
Playing time (max.)	74 minutes (officially, but a few CDs manage to go a little over this duration)
Total track length	5300 m (about $3^1/_4$ miles!)

18.5 Other digital recording systems

Before closing this chapter we should make a brief mention of two digital disc systems.

1. **Recordable CD** (CDR): These are being used increasingly in professional studios. These look very much like conventional CDs and can be played on a standard CD machine. However they are bought as blanks. The

recording is made using a laser to alter the characteristics of a special dye. The resulting colour change is equivalent to the pits on a normal CD. A CDR is a record-once disc, which means that it cannot be erased and recorded on again, although it can, of course, be replayed as many times as a standard CD.

2. **Magneto-optical discs**: In these there is a special plastic layer and this is heated by a laser in the presence of a magnetic field. Unlike CDRs they can be re-recorded.

19

More digital devices

We've seen that some very clever recording operations are possible with digital signals. However the story doesn't end there. Since, in any particular system, the correct 1s and 0s are all the same size and occur at specified time intervals it becomes possible to manipulate the *bit stream* – the series of 1s and 0s – in a variety of ways and still restore things to the original pattern when necessary. Further, bits can easily be stored in computer-type memories: for many purposes a temporary storage only is needed and this can be provided by RAM – *random access Memory*. The following is a brief account of some of the more important applications of digits.

19.1 Artificial reverberation

Artificial reverberation (often called echo, although this is not strictly correct) is a most useful addition to many sounds. It can be used to enhance a too-short reverberation time (r.t.) for music; it can help to give realism in drama, whether for radio or for television, where a scene is supposed, for example to take place in a church but the r.t. of the studio is far to short; it can be used to alter sound perspectives, and so on.

Originally artificial reverberation was provided by a bare room containing a loudspeaker and microphones. The latter picked up the sound which had bounced round the bare room and their output was mixed in with the original signal. *Echo chambers*, as they were called, were expensive in terms of space and the r.t. couldn't easily be altered. Later came the *echo plate*, a sheet of steel about 2 m by 1 m, suspended vertically and under tension. Fitted to it is an input transducer rather like the transducer of a loudspeaker. This sets the plate into vibrations which are reflected from the edges of the plate. It's rather like a two-dimensional room. Other transducers pick up the reverberations. The r.t. can be varied by moving a *damping plate* nearer or further from the steel plate. Echo plates can give quite good artificial reverberation but they sometimes sound 'tinny'. However they are fairly large and have to be mounted away from the studio or its control room because loud noises in their vicinity can be picked up by the plate.

Another analogue device was the *echo spring*. This consisted of a pair of metal springs which were made to vibrate by an input transducer at one end and a pickup transducer at the other end. Two springs of different characteristics were used to give a stereo output and also to introduce a slight element of randomness. Spring devices could be cheap and compact (not all were) but they tended to have a 'twangy' quality. The bigger and more expensive managed to avoid this, though.

The picture changes with the use of digits! The actual circuitry and logic behind a digital reverberation device is very complex. It must be enough here to say that the digital signal is fed into a large number of memory stores. Samples are taken out of the stores in a way which simulates the decay of a reverberant sound, so that the amplitude decreases with time. There has to be a degree of randomness for the final analogue sound to be convincing. However digital devices can provide excellent reverberation and the r.t. can be varied over very wide limits. Not only that but the *initial time delay* – the time gap before the start of the reverberation – can be varied. (This is an important characteristic of real-life reverberant conditions.) Also the r.t. can be varied with frequency, again a feature of real reverberation. And all of this can be contained in a compact unit and at a cost, in real terms, which is far less than that of an echo plate. In fact there are very satisfactory digital reverberation units which cost little more than a respectable domestic CD player.

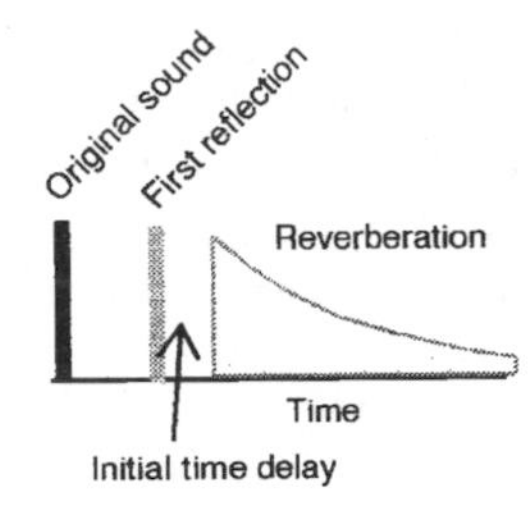

If the initial time delay (i.t.d.) is more than about 30–40 ms then the brain is aware of this as a perceptible time gap and we call it an 'echo'. If the i.t.d. is less than 30–40 ms we are not consciously aware of it but we gain an impression of space

19.2 Pitch change

It becomes quite easy (in principle!) with digits to change the pitch of a sound. First, let's remind ourselves that when we change the tape speed on an analogue recorder we can alter the pitch of the sound but the change in speed may not be what's wanted. It's sometimes useful to be able to change the pitch but without altering the speed, and this is where digital memories (stores) come in. Figure 19.1 tries to show the idea. The reader should note that it bears absolutely *no* resemblance to the way

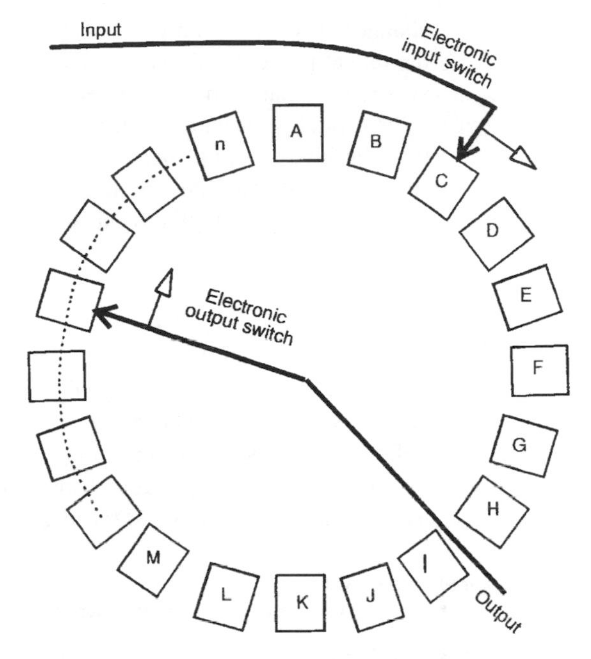

Figure 19.1 The principle of pitch changing

in which real pitch changes are made. It simply illustrates the way in which they work.

A, B, C. . . to *n* are memory stores. Imagine that the electronic Input switch rotates round the stores in a clockwise direction and puts a digital sample into each store. The contents of each store are read by the electronic output switch, which also rotates clockwise.

Now if both switches rotate at the same speed then there is just a time delay between input and output. If, though, the output switch rotates faster than the input then samples are being taken out at a faster rate than they are put in, and when they're converted back into analogue signals we have an *increase* in the pitch.

But, the reader may ask, what happens when the output switch catches up with the input? It has then emptied all the stores behind it and the stores in front of it haven't yet been filled with new samples! When this happens the output switch is made to read repeated stored samples. This can cause an audible effect, known as a *glitch*, and in the earlier pitch change devices glitches could be objectionable unless the change in pitch were very small, so that the glitches were relatively infrequent. Modern pitch changers get round the problem very neatly by arranging that the repeat process only happens when the audio signal level is very low, so that silence, or something near to it is being repeated.

The pitch is lowered when the output switch goes slower than the input. Again, there can be a glitch when it has lagged sufficiently to coincide with the input, and then the stores have to be emptied rapidly, but if this takes place at low audio signal levels the effects are not disturbing.

19.3 Time division multiplex (TDM)

Digital pulses can be compressed in time and restored to their original spacing later. This makes it possible to send a number of different channels of data along one signal path. Because a unit of time is shared between many channels the process is known as *time division multiplex*, or TDM. Figure 19.2 illustrates this.

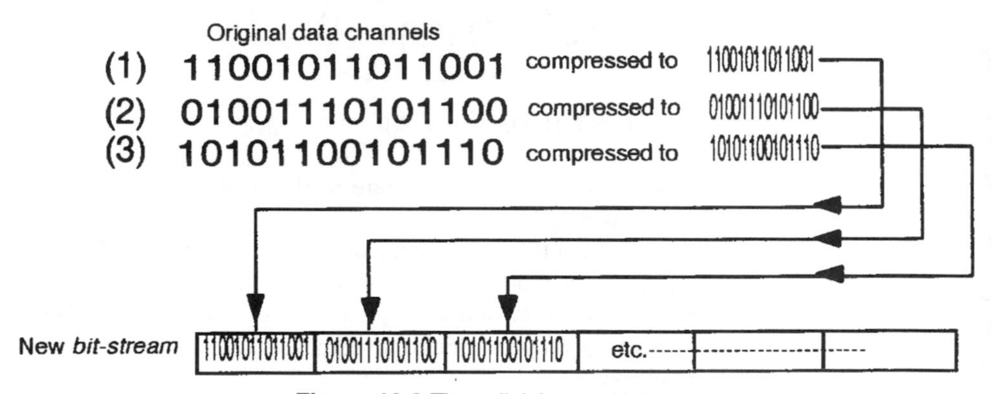

Figure 19.2 Time division multiplex

(1), (2) and (3) are digital samples (14-bit in this case, for simplicity). Each is compressed in time and then put in sequence into the final bit-stream. A more practical example is shown in Figure 19.3, which represents in simplified form the distribution system used by the BBC for sending its radio programmes around the country from studio centres to transmitters.

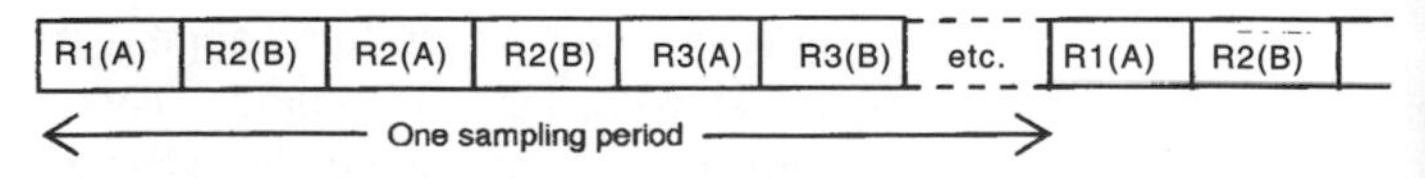

Figure 19.3 BBC radio distribution system (simplified)

R1(A) represents a compressed sample of Radio 1, stereo left (A), and so on. We have only shown Radios 1, 2 and 3, but successive blocks would represent the remaining radio channels and also various internal data channels, making about 18 in total. All of these *must* be contained in one period of the original sample. Then we start all over again with R1(A). At the destinations the wanted programme samples are extracted using a timing system and expanded back to their original form.

All in all a TDM system such as this makes for great economies: instead of having, for example, 18 different data channels using 18 separate pairs of wire, only one is needed.

19.4 Sound-in-syncs (S.i.S.)

Up to around 1970 television sound was sent from the studio centres to the transmitters along one route (cable or other kind of link) and the vision signal along a separate route. The s.i.s. system was developed so that both were sent along the same route. This reduced costs and also increased the reliability. It also meant that it was much more difficult for mistakes to be made at switching centres resulting in the sound for one channel appearing at the viewers' television sets accompanied by pictures for a different channel!

Basically the idea is that the digitized sound signal is tucked into the area of the vision waveform normally occupied by a synchronizing signal, as Figure 19.4 shows.

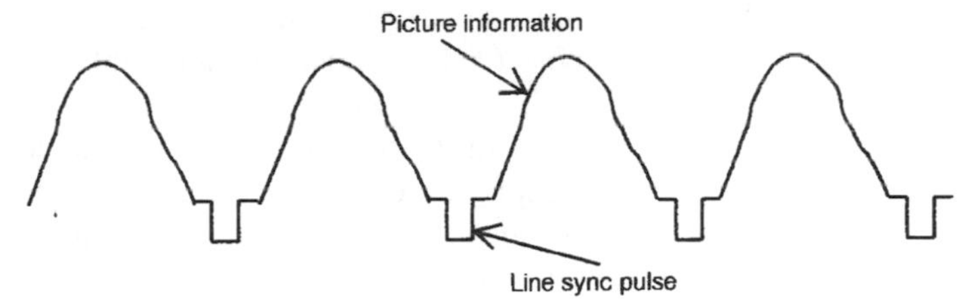

Figure 19.4 Television waveform

The process is actually rather complicated. The digital audio is processed to a 10-bit sample, which may not seem very good, but there is a companding process which makes the overall result more like a 12-bit system. (This may seem to be very inferior, but in fact is perfectly satisfactory, remembering that transmitters and receivers are bound to cause some impairments; for example at the transmitters the digital audio is converted back to analogue

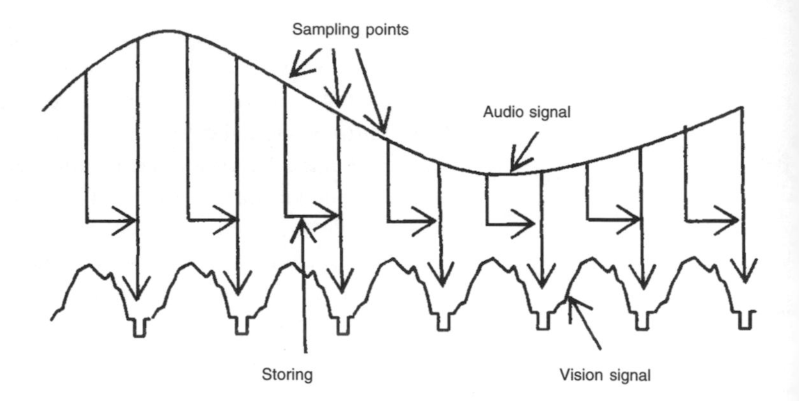

Figure 19.5 Slotting audio samples in the vision signal

for transmission. This has changed with the advent of NICAM stereo for television, which we shall deal with later in this chapter.)

The sampling rate for S.i.S is 31 250/s. At first sight this seems a very peculiar number but it is, in fact, twice the line frequency of a 625 line television system. In other words the audio signal is sampled twice in each line, once at the end of the line and once at the middle of the line. The mid-line samples are then stored and slotted with the end-of-line samples into the sync region, as in Figure 19.5.

There is more manipulation than this. One of the samples is inverted: the 1s become 0s and the 0s become 1s, and then this inverted sample is interleaved with the non-inverted one, the object being to maintain a roughly constant d.c. equivalent.

Then this sequence is turned back-to-front so that the most rapid fluctuations which will come normally at the end of the sample now come at the start. This is to reduce the likelihood of fluctuations disturbing the next part of the video signal. Figure 19.6 summarizes these processes.

All this is rather complicated and the reader need not try to understand it all: we mention it simply to show the kind of operations that can be carried out with digital signals. S.i.S. is a very good example. Perhaps we should emphasize that it is used in the UK only for carrying television signals to the transmitters. There the normal sync pulses are restored, otherwise domestic receivers would probably have great difficulty in producing satisfactory pictures.

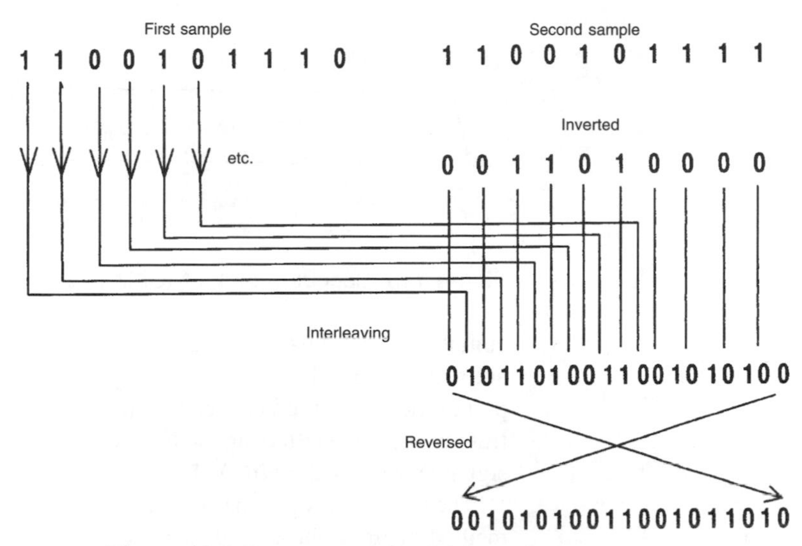

Figure 19.6 Summary of the S.i.S. operations

19.5 NICAM 728

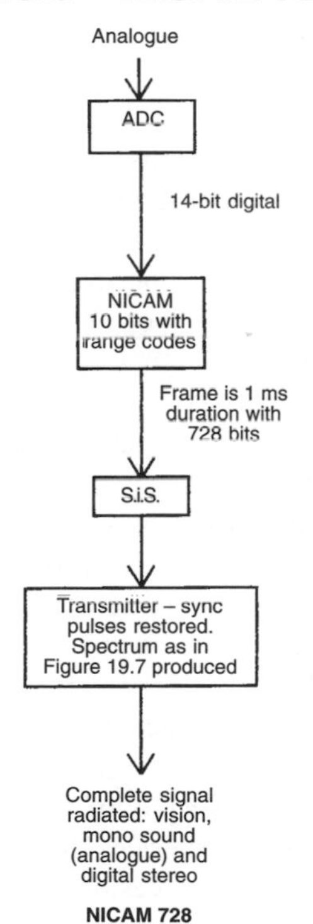

With the advent of stereo sound for television there had to be a way of delivering two high-quality audio signals as well as the picture information to the domestic receiver and NICAM 728 does this.

The reader may be getting confused over the terminology! We've already outlined one NICAM when we were looking at methods of reducing bit rates. NICAM 728 is a development from it. The trouble arises because the domestic television set industry, in using the new NICAM 728 has chosen to drop the 728, presumably because it might confuse the public. Maybe, but in doing so it may tend to confuse people like the readers of this book. Here, to avoid this situation we shall stick to NICAM 728. The reader then only needs to know that a television set, labelled 'NICAM Stereo Sound', implies NICAM 728.

NICAM 728 is a development of s.i.s. in that the sampling frequency is the same – 32 kHz. The resolution initially is 14 bits, but this is reduced to 10 bits, with five ranges – in other words using the first NICAM process. The digital signals, of which there are two lots for stereo, together with other data such as parity bits and range codes, form a *frame* with a duration of 1 ms and each frame then contains *728 bits* – hence the name.

All this information is carried to the transmitters tucked into the line sync period, just like ordinary S.i.S, but then a rather

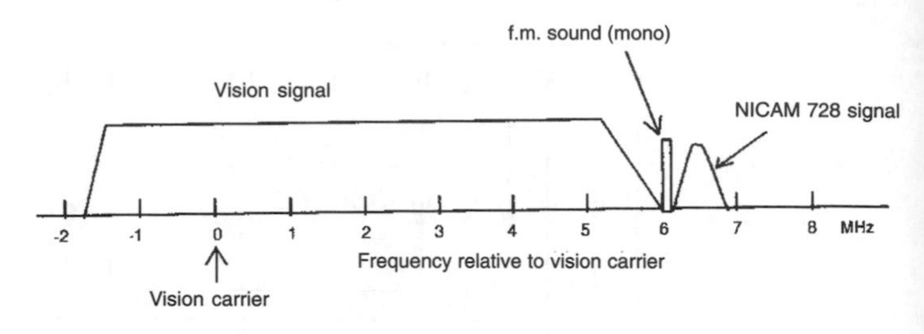

Figure 19.7 Spectrum of the NICAM 728 signal

complicated modulation process is carried out. The line syncs are restored and the NICAM 728 signals are used to modulate part of the transmitted carrier. Figure 19.7 shows the basic spectrum of the transmitted signal. Notice that there is an f.m. sound signal as well as the NICAM 728.

The f.m. sound signal has for a very long time been the standard method of transmitting television sound and contains the mono version for those receivers not equipped with NICAM 728.

19.6 Time code

This is primarily a means of providing accurate synchronization between two or more tape machines, whether they are audio or video. Even a very high-class analogue audio tape machine cannot give really accurate timing. One cause of this is slippage of the tape as it goes past the capstan, so that there can be a timing error over half an hour. This may be only a few seconds but enough to mean that synchronization with other machines is impossible. The problem is got round by using one track for digital timing information, known as *time code*.

This extra track can mean loss of a track that could be used for audio or video signals, but in fact this isn't as serious as it sounds. We mentioned in connection with multi-track tape machines that time code was usually put on to an outside track, and this can still leave plenty of audio tracks. Video recorders usually have an extra track for this purpose, and remember that being digital means that time code doesn't need to be a very accurately recorded waveform. And there is a very ingenious way of having the time code signals on a centre track on a stereo machine with only two tracks.

Time code was originally devised by the Society of Motion Picture and Television Engineers (SMPTE) in America. The code was then adopted in Europe by the European Broadcasting Union (EBU) It's full and correct title then is *SMPTE/EBU time code*.

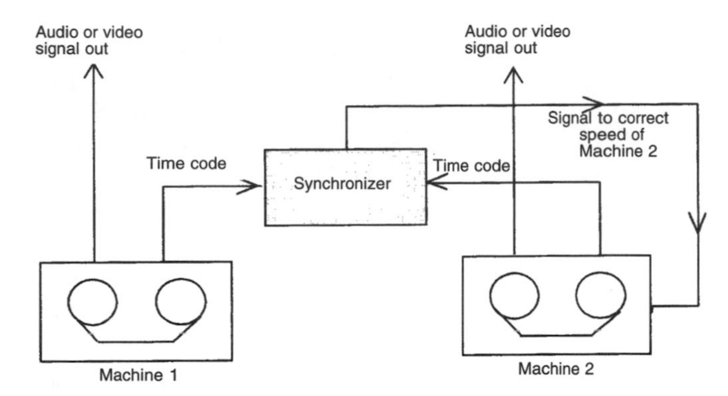

Figure 19.8 One use of time code

However it's usual to simplify this and refer to it as 'SMPTE (pronounced *simpty*) time code'.

The digital signal is made up of 80 bits which contain information about the time in hours (usually starting from the beginning of the tape, although clock time can be used), minutes, seconds and *frames*, which in the UK and most of Europe are twenty-fifths of a second.

It's normal to put the time code on to the tape before it's used, a process called *striping*, using a *time code generator* which may be built into the mixing console, or may be a separate portable unit.

Then, to keep different machines in synchronization, which means of course that all the tapes must be already striped, a *synchronizer* is used, as in Figure 19.8.

Time code can be used for much more than just keeping tape machines in 'sync'. It may be very useful where two pieces of audio that need to be in sync have different time codes recorded on them. Then one machine can be run with an *offset* – deliberately behind or in front of the other machine(s), and the time code synchronizer can be set to do this. Time code can also be displayed on a picture monitor and this allows very accurate editing of pictures and their sound accompaniment. Yet again, tape machines can be put under the control of a computer which can locate points on the tape quickly and with great accuracy. In short, time code is a very valuable aid to all sorts of sound and vision manipulations.

19.7 The AES/EBU interface

This sounds heavy stuff! However, all we need to do here is to let the reader be aware of its existence.

With so many digital devices around, often made by different manufacturers, there needs to be a standard means of interconnecting them. Some manufacturers have their own *interfacing* systems – obviously compatible with their own equipment – but there's obviously a need for a universal one as well.

The AES/EBU interface is just such a means of connection. (AES is The Audio Engineering Society, an American-based organization but with branches in almost all other countries. EBU we have already mentioned as standing for the European Broadcasting Union. The AES/EBU system is therefore acceptable in the USA and also Europe, but many other countries, for example in Asia, also accept it.)

Without going into great detail we will simply summarize things here by saying that the AES/EBU interface allows two digital audio channels to be sent along one cable for distances of up to about 100 m – more in certain circumstances. This may not seem very far but the object is to provide links between equipment within a studio complex.

19.8 MIDI

This stands for *musical instrument digital interface*. It's an internationally agreed standard for linking together electronic musical instruments with each other, or with computers and other devices capable of producing music.

There is no such thing as a particular MIDI circuit – instead it's a recognized way of formatting the necessary information. Different manufacturers may have quite different ways of producing the wanted effects. These include things like the pitch of the original note and its duration, the velocity with which the key was struck, and so on. Thus a simple keyboard which is equipped with MIDI can control synthesizers, samplers and a variety of music- or sound-producing pieces of equipment.

19.9 ISDN

The *integrated services digital network*. This is a kind of extension to the telephone system (BT in the UK). It makes it possible, say, for a news reporter to make a speech contribution into a programme. A special unit is needed. The microphone is plugged into this and there is a connection to a telephone socket. Two fairly high grade circuits plus a 'control' circuit can operate in each direction.

20

Public address

20.1 A definition

The term 'public address' (PA) means what it says, except that there are actually at least two quite different ways in which the 'public' may be addressed through loudspeakers. One is the sort of thing found at, for example, race meetings or railway stations. There is a large area to cover, the microphones are well away from the loudspeakers, probably in an enclosed room some distance away, and intelligibility rather than sound quality is the main requirement. This form of PA isn't often of direct concern to broadcasters. It may well be heard in the background and thus can be a useful way of conveying something of the atmosphere of the event or location. Beyond that, though, it is rarely of great significance.

The second PA situation is the one where the microphones and the audience are in the same room (a studio) and reasonably good quality is of importance. Very often the need here is really for *sound reinforcement*: the audience will be likely to hear some of the programme sound but it will need a degree of amplification for it to be clearly heard. This is the form of PA that most often concerns broadcasters and it can frequently be a problem. So we shall concentrate on this aspect of PA. 'Sound reinforcement' would generally be a more accurate term to use, but rightly or wrongly, 'public address', or 'PA' for short, has come to stay.

In this chapter we shall also deal with a very specialized form of PA called *foldback*, FB for short, but we'll come to that later.

20.2 Why is PA needed?

There are many reasons why some kind of PA is often wanted in a studio. The first point, an obvious one, is that PA is only needed when there is a studio audience. So why have studio audiences? Let's try to answer that one first.

The following are typical situations where a studio audience is desirable:

1. Situation comedies, in either radio or television. An audience which laughs in the right places can give the actors something to play to, especially if they are used to working in the theatre. Such an audience can, in some difficult-to-define way, add to the enjoyment and fun of the listeners and viewers at home.
2. Stand-up comedians almost invariably want an audience to perform to.
3. Musical groups, whatever the kind of music, can benefit from the presence of an audience.

With the exception of orchestral music some PA is almost invariably needed for each of these situations. A particularly good example is in television 'sit-coms'. Here the acting generally needs to appear to be fairly natural, with speech levels comparable with those used in everyday life. A studio audience would find this difficult to hear, and react to, without PA. And in the case of a band with, say a vocalist, the singer is very likely to be inaudible against the instruments unless PA is provided; also some of the weaker instruments such as acoustic guitars need amplification.

An obvious problem is that the PA is generally going to need to be selective; in a band for instance the vocalist will need much more amplification than, say, an acoustic guitar. The audience will be likely to hear the louder instruments perfectly well, so that the *PA feed* to the loudspeakers needs to be *balanced* if the audience is to hear something reasonably close to what the listeners and viewers at home will hear.

The reader will quickly appreciate that there are several basic requirements for studio PA. We can list the most important ones as:

1. The audience must be able to hear an appropriate version of the studio output.
2. There should be no regions where the PA level is either too loud or too quiet.
3. The studio microphones should not normally pick up any of the PA, causing *colouration*. By this we mean the vaguely echoey sound that immediately betrays the presence of a PA system. Such colouration would be disastrous in a sit-com as it would immediately destroy any illusion of reality. (There may be some occasions where a little PA colouration might be appropriate in order to suggest a particular environment but these are rare.)

4. Most important, there must be no squeals caused by the microphones picking up enough of the PA to feed it back into the PA system. Such noises are called *howl-rounds* or *feedback*. The former term is perhaps to be preferred as it describes fairly accurately what is happening, and the term 'feedback' can mean different things in other circumstances. We will look at the howl-round problem first.

20.3 The cause of a howl-round

Figure 20.1 represents the basics of a PA situation. It will be seen that there is a kind of loop made up of the loudspeaker (in practice, of course, there'll usually be more than one), the path of the sound waves from it to the microphone(s), shown here as a straight line but really it will be a complex path madc up of many reflections from walls, floor and so on, and then from the microphone through the amplifier driving the loudspeaker, and back to the loudspeaker.

Now it may be that the attenuation in the path through the air, together with losses in the microphone and loudspeaker, totals *less* than the amplification, in which case there's no problem. We would say that the *loop gain is less than unity*. But, if the

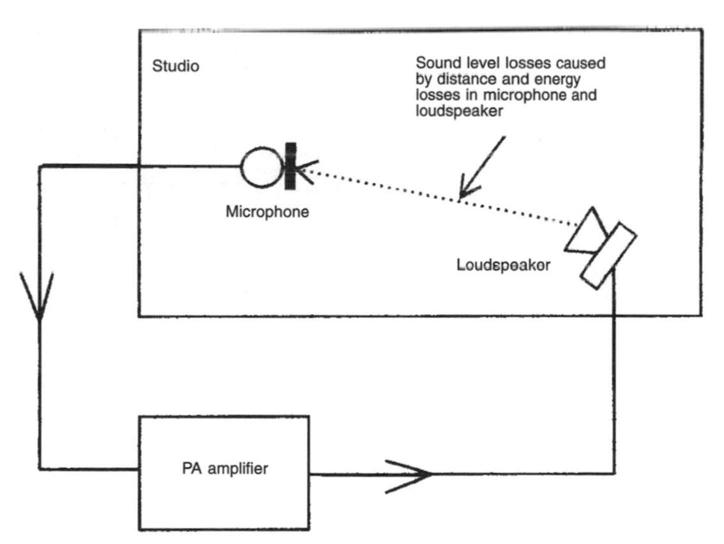

Figure 20.1 The basics of a PA situation

gain of the amplifier is greater than the combined losses, making the loop gain greater than unity, then there's an instability and oscillation, in the form of a howl-round, will occur. (The concept of loop gain or loss is quite an important one and has wide applications.)

It might seem, then, as if all we have to do is to keep the amplifier gain setting sufficiently low and all will be well. Unfortunately it's not quite as simple as that. Without going into too much detail let's see what the complications are. Figure 20.2 shows, in a rather idealized form, the characteristics of the main components in the chain.

The diagram shows three frequency responses: at the top the sort of thing one might get if the response of the studio between loudspeaker and microphone were measured. This typically, but of course not invariably, might have peaks and troughs of up to about 10 dB above and below the mean. These fluctuations are a consequence of standing waves, and we should remember that

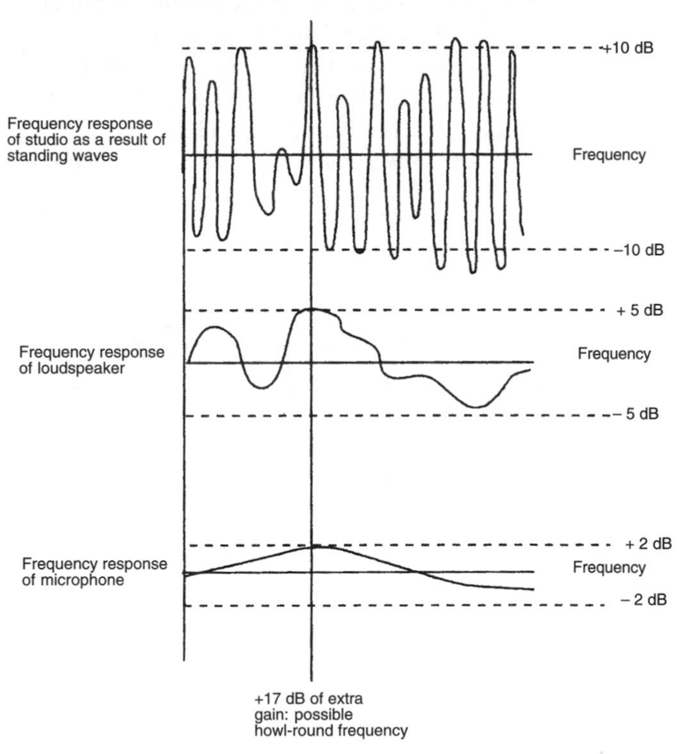

Figure 20.2 Conditions for a possible howl-round

the frequency response will be different if we have either the loudspeaker or the microphone in different positions, or if the studio contents are changed, as for example different sets in a television studio.

The second graph shows the frequency response of the loudspeaker – again not perhaps very realistic but nevertheless perfectly representative in that there could be variations of 5 dB above or below the mean. (Many PA loudspeakers may not be even as good as this!)

And finally the microphone, which will probably have the flattest response of all, but even so is unlikely to be perfectly flat.

Notice that we've drawn a vertical line at a particular point where there happens to be a 10 dB peak in the studio response, a 5 dB peak in the loudspeaker's response and a 2 dB rise in the response of the microphone, making a total of 17 dB *to be added to the loop gain*! This is the frequency at which a howl is most likely. Turning the amplifier gain down considerably may well stop the howl but the level of the PA for the audience may then be too low for reasonable audibility. Luckily there are solutions to the problem, albeit none of them perfect, which we can now look at.

20.4 Reducing the risk of a howl-round

A PA squeal is embarrassing if it occurs at the village fete but it is a really serious matter in the professional world. To make matters worse there's no sure-fire way of ensuring that howl-rounds are not going to occur in any particular set of circumstances. One thing is most important, though: there must be thorough checking of the system before the recording or transmission! If this is done then the likelihood of trouble later is obviously very greatly reduced.

In broadcasting there are broadly two kinds of PA set-up: permanent ones, which are a feature of most large production studios, and temporary ones at, for example, an outside broadcast (OB) venue. In either case the following guidelines can be useful:

1. The right microphones in the right places. This is not perhaps as easy as it might seem. Directional microphones, cardioids for instance, with their insensitive side(s) towards the PA loudspeakers can obviously help, but this is not always a complete answer. PA sound can be reflected off a wall on to the microphones and this makes any directional properties rather less useful – see

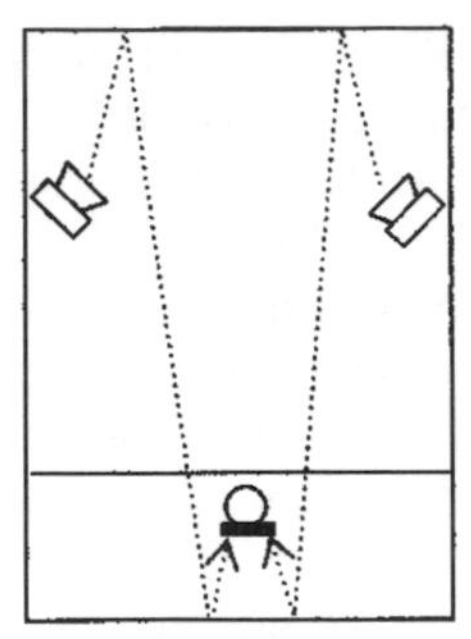

When directional loudspeakers may not be a great help

the small diagram in the margin.

At first sight it would seem that omnidirectional microphones are to be avoided. But, interestingly, the small omnidirectional personal microphones worn by presenters can often work surprisingly well when there is PA.

Then the siting of microphones can be very important. And it should be remembered that the frequency response of the studio, as in Figure 20.2, will change if the microphone is moved. So, a microphone in one position – on a stand for example at the front of a platform – may be perfectly all right, but if it's moved by perhaps only a matter of centimetres there could be a howl. And this is an excellent reason for checking thoroughly in advance. If a microphone is to be moved it must be taken into all the likely (and even some of the unlikely!) places at the rehearsal.

2. The type and siting of loudspeakers. Here there almost always have to be compromises. The obvious kind of loudspeaker to have for PA work is the column or line-source loudspeaker which we looked at in Chapter 11. It's possible to have almost complete freedom from howl-rounds if the microphone is directly below the loudspeaker. Unfortunately a loudspeaker directly over the microphone is not generally feasible in broadcasting! It's likely to be visually quite unacceptable in television, and apart from that there may be great difficulties in rigging the loudspeaker in that position. Nevertheless there can be quite satisfactory results from having line-source loudspeakers slung above a studio audience and with their 'dead' directions pointing towards any microphones.

Having said that it isn't normally feasible to have microphones in line with the 'dead' axis of a line source loudspeaker, there is an exception to that. With a studio audience that is there for the purpose of providing laughter there need to be *audience reaction microphones*. There's no point in having a laughing audience if the laughter isn't heard by the listener or viewer! Microphones are therefore generally slung over the audience to pick up the audience's reaction, and these mustn't be anywhere where they can pick up the PA loudspeakers' output. Directional microphones close to

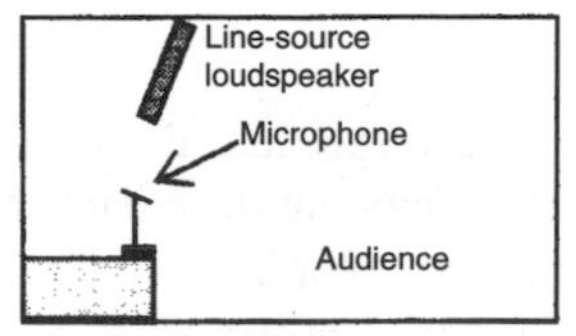

An almost perfect, if not often very practical, PA set-up

the ends of line-source loudspeakers can be a very satisfactory answer.

However it's an unfortunate fact of life that the sound quality from line-source loudspeakers is usually not very good. It can be perfectly adequate for speech, but not good enough for music PA. If high quality PA is needed then it is possible to use non-directional high grade loudspeakers without as much trouble as one might expect. The reason is that such loudspeakers have a much flatter frequency response than line-source ones, and if there are fewer marked peaks then the situation shown in Figure 20.2 is just a little bit better.

3. Electronic devices. There are two approaches to reducing PA howlrounds. One uses graphic equalizers. The howl frequency is located and then the appropriate channel on the equalizer is reduced in level. It's best not to reduce the level by more than is necessary as the impairment of the sound quality becomes too noticeable.

   The other electronic device is a *frequency shifter*. By moving the entire audio spectrum fed to the loudspeakers by a small amount – a few hertz – the peaks in the studio frequency response are, as it were, moved along slightly and there can be cancellation in the total loop because peaks now occur where there were troughs. These things have to be used with great caution, as if the audience can hear some direct sound as well as the PA sound there can be a disconcerting effect. The effect would be even worse if a singer were to hear PA sound which has been frequency shifted!

4. A PA mixer. By this we mean a person doing a separate balance of the PA. This means a skilled member of the sound crew sitting in the audience area with a mixing desk. This is fed with selected outputs from the main mixer and the PA is then kept firmly under control. Not only can the PA quality be better but also potential howl situations can be anticipated. The drawbacks are that expensive extra equipment is needed and a skilled operator is not available for other work.

To summarize this section we hope that the reader will have gained the impression that PA in a studio can be a real headache for the audio staff!

20.5 Foldback (FB)

We said earlier in the chapter that this was a specialized form of PA, and so it is. However there's a very important difference, at least as understood in the broadcasting and recording world. In essence: *PA is for the audience FB is for the performers.*

There are many occasions when the artists need to be fed with some kind of sound. Typical examples are:

1. Actors in a play may need to hear a pre-recorded sound effect in order to react to it.
2. A singer might need to hear a pre-recorded music track or an orchestra. (In television work picture monitors visible to the singer may be necessary, but these and the FB loudspeakers must either not been seen by the cameras, or they must be 'disguised'. In sound-only work headphones can be used instead of loudspeakers.)

The problems of foldback can be the same as those for PA if there are microphones in the same studio as the loudspeakers, but of course this doesn't matter if the foldback sound has been pre-recorded. In television a typical place for the FB loudspeakers (which shouldn't be visually intrusive) is on the studio floor in front of the performers. They can sometimes be made to look like part of the set.

In the next chapter, dealing with sound mixing consoles, we shall see that the means of providing the audio signals for both PA and foldback are essentially very similar.

20.6 Safety

This is an appropriate place to emphasize the absolute necessity of safety. We'll return to this matter later, but with PA it's probable – perhaps almost certain – that there are going to be loudspeakers (and audience reaction microphones) positioned over the heads of members of the public, artistes and members of the studio crew. *These must never be a risk to anyone.*

Mountings must be secure; *safety chains* – additional strong chains looped round a suitable structure – should be used so that even if the main mounting were to fail the chain would prevent the microphone or loudspeaker from falling on anyone.

And not only that, there must be the greatest care taken when rigging these things. No one should be standing directly underneath. If the device is heavy then one person on his or her own should not be trying to manipulate it.

Safety must come first.

21

Sound desks (mixing consoles)

21.1 What is a sound desk?

In this book we shall use the term sound desk (alternative terms for the same thing are *mixer*, *mixing desk* and *mixing console*) to mean a unit whose primary functions are to:

1. Provide amplification for the weak signals from microphones and other sources.
2. Allow an operator to control the level of such sources and mix and balance them, so that the relative loudnesses of these sources in the output are artistically satisfactory and within the required technical limits (for example avoiding overloads).
3. Provide monitoring so that sources and outputs can be checked both visually and aurally.
4. Provide communication facilities with other staff in the studio and other, possibly remote, studios when necessary.
5. Allow the audio signals to be processed in the ways we have seen in an earlier chapter, add artificial reverberation and provide means of connecting additional devices into the system.
6. Provide the means of selecting appropriate outputs for studio public address (PA) and foldback (FB).
7. Provide phantom power at microphone sockets.

The size and complexity of a sound desk can be roughly indicated by the number of *channels* it can accommodate. By 'channel' we mean in this context the number of paths for individual sources. A channel would typically accept the output from a microphone or other source and provide the facilities for processing that signal, selecting it for PA or FB and containing a *fader* for accurate adjustment of the signal level. (All this may seem rather fearsome, but the reader will find that it will become much clearer as we go on.)

Returning to the size of a sound desk in terms of the number of channels we can say that a small portable one might have no more than four channels. (In such a unit the term 'sound desk' is a bit pretentious so we'd simply call it a 'mixer'. However its basic structure in signal handling terms will be very little different from a large desk. The basic skeleton of a mouse isn't really very different from that of an elephant – both have legs, backbones, skulls, ribs, and so on!)

A large sound desk in the broadcast and recording world can have anything up to *100 channels*, and maybe more. Such sound desks are very complicated to look at but we should remember that if there are, say, 100 channels, then these are all going to be similar and we need only study one of them!

What we are going to do is to look at the structure of a typical medium-sized sound desk, taking it in very easy steps. While there are going to be differences between any two makes of desk and to study them all would be impossible, nevertheless we can, in looking at a typical one, go a very long way towards seeing how they all function and their controls and other facilities are related. So let's start with a typical channel.

21.2 The facilities of a typical channel

General symbol for a jack or any other audio socket

An arrow means variability. This is an amplifier whose gain is varied by a control knob or switch

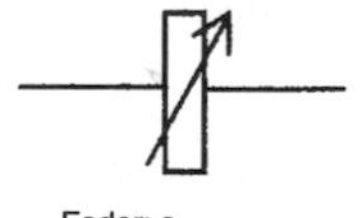

Fader: a variable resistor

What we are going to do is to show the workings of the main parts of the desk in the form of *block diagrams*. This means using symbols to represent pieces of circuitry. We've already done this earlier in the book by using a triangle on its side to represent an amplifier. It doesn't matter how simple or how complicated the actual circuit is – we just need to know that it's an amplifier.

The ability to follow a block diagram is more-or-less essential if one is ever going to make use of its facilities fully, and while this can be a little tricky sometimes (depending on how well laid out the diagram is on paper) it isn't really all that difficult.

Figure 21.1 shows a rather simple sound desk channel – we shall add things to it a little later. The reader will recognize some of the symbols and probably guess at the meaning of some of the others. Diagrams in the margin can clarify the rest.

Important: In all the diagrams in this chapter we are omitting the many little amplifiers which are likely to be present in a real desk. Some of these will be for raising signal levels after losses have occurred; some will serve as isolating devices to prevent

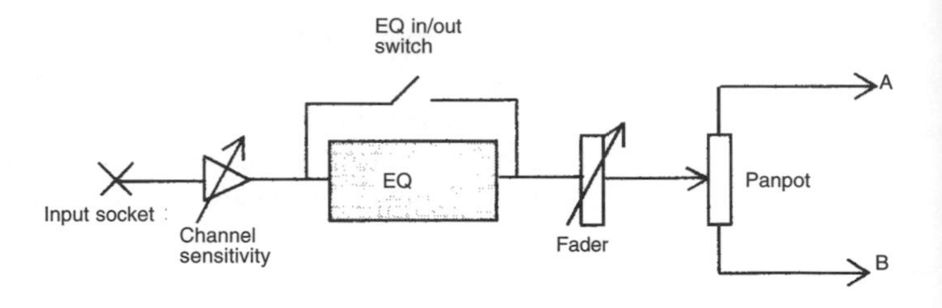

Figure 21.1 A very simple sound desk

unwanted interactions, and so on. They're all going to be necessary but they aren't operational controls, so we are leaving them out for the sake of clarity.

We would hope that Figure 21.1 is pretty self-explanatory, with the possible exception of the shaded box. EQ simply means that this represents circuits for providing top lift and cut, bass lift and cut and presence. Depending on the size (and cost) of the desk it might be more or less than this. Now it's often very important to be able to switch the selected EQ in or out; for example at rehearsal to compare with the original and check that the wanted effects are being produced. The switch above the box, when it's open, as it is in the diagram, means that the audio signal is passing through the EQ box and is therefore available for manipulation. If the switch is closed the EQ box is effectively short circuited and therefore has no effect.

Figure 21.1 is a very simple diagram and it doesn't show any provision for:

1. Coping with relatively high level signals, maybe approaching zero level from, say, CD players and tape machines, which could overload the amplifiers handling microphone level signals.
2. Doing anything about possible phase reversals.
3. Allowing other equipment, compressor/limiters or graphic equalizers, for example, to be inserted into the channel.

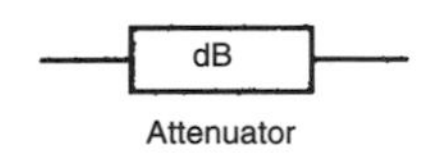

Let's deal with these one at a time and then incorporate them into a complete channel diagram:

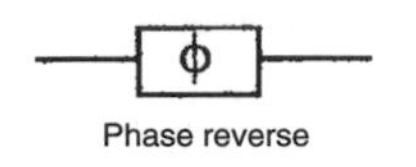

1. Coping with some sources which are high level ones. There are in fact various ways of dealing with this. The one we'll illustrate is to have *two* input sockets, one of which would be called the *high level input* and the other,

naturally, the *low level input*. The are linked with a little box with 'dB' inside it, this indicating that it is an *attenuator*. Sometimes the number of decibels of attenuation may be stated outside the box.

2. A unit for reversing the phase of the signal – and remember that a single line in a block diagram represents *two* wires. It might be nothing more than terminals at which the wires are crossed over. A box with the Greek letter ϕ ('phi', pronounced *fie*) inside can represent this. And if there is a switch outside then this implies that the phase reverse can be switched in or out.

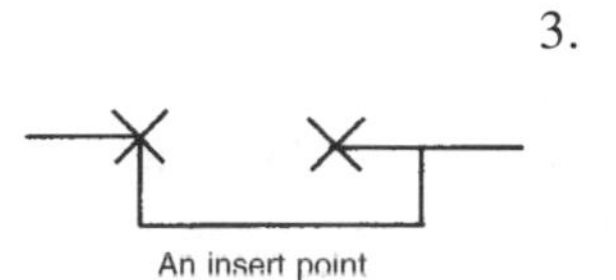
An insert point

3. To allow for other 'outboard' equipment to be inserted into the chain we have an *insert point*. It's an accepted convention that if a line comes out of a jack symbol vertically, either going up or down, it implies that this is a *break jack*. That is, inserting a plug *breaks* the connection from the main route and the audio signal now flows into the plug only. An insert point then has two jacks so that the input to the extra device is plugged into the left (break) jack and its output is plugged into the right jack.

Putting all these things together we now have Figure 21.2.

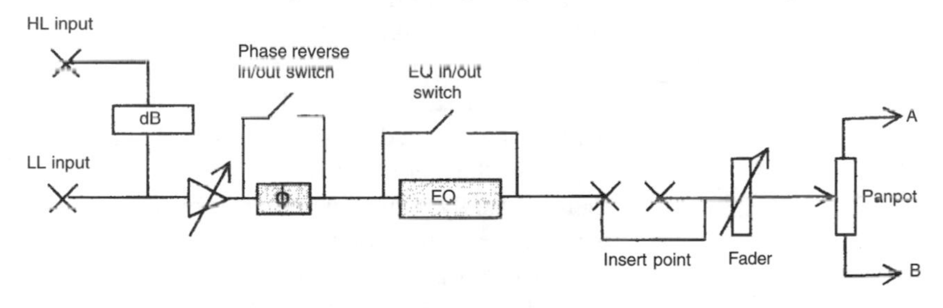

Figure 21.2 A more realistic channel

It's just the earlier one but with additions. At this point we must say that the actual order of things can vary from one type of desk to another: the phase reverse might come before the channel sensitivity, for example. Also a large desk may have insert points both before and after the EQ unit.

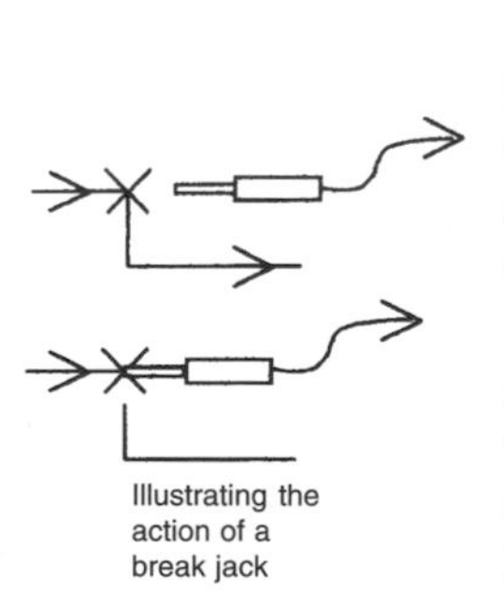
Illustrating the action of a break jack

Next we have to think about *auxiliary outputs*, which sounds new but in fact means nothing more than those audio signals which are to be sent to things like artificial reverberation units (which we shall call echo – incorrect but often used!), PA and FB. There are many places where these signals could be derived

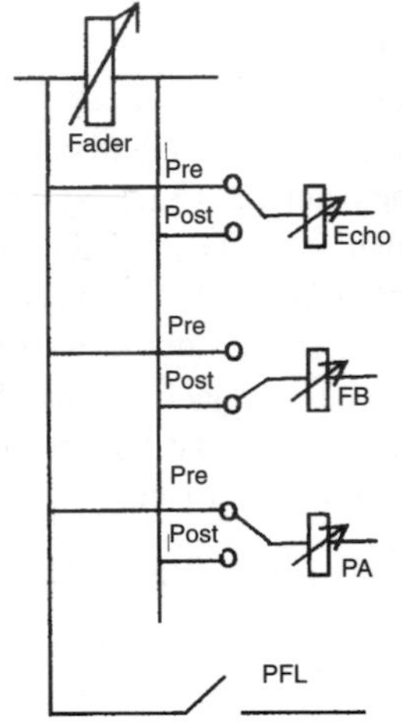

from the main signal path, and a good question is whether this derivation should be before or after the channel fader. Taking PA as an example, if the PA feed is taken before the fader then the PA signal is constant regardless of the fader setting. In some situations this could be a good thing. If it is taken off after the fader then the PA level will vary with the fader setting and in some circumstances this could also be useful. The solution adopted in many desks is to allow the operator to choose which he or she wants: what is called *pre-fader operation* or *post-fader operation*. Our margin diagram shows how. Note that each feed has its own little fader – in reality a fairly simple knob working a volume control.

There is one further item on the diagram which needs explaining; it's the output labelled PFL. This stands for *pre-fade listen*, sometimes called *pre-hear* (PH). What this does is to enable the operator to check on the quality and level of a source before fading it up. This can be particularly important when working with a remote studio. If quality and/or level are wrong then it's to late if this isn't found out until the contribution has been mixed in with the rest of the programme! PFL (PH) gets round this situation.

It may be appropriate here to mention that there are other types of specialized checks like PFL. Without adding them to diagrams which might make these look very complicated we will just mention three of them:

After fade listen (AFL) lets the operator single out a source when it is already part of the programme sound to check its quality. The selected source may be fed to another loudspeaker, or put on the main loudspeakers with all other sources cut. There's usually a little warning light near the meters to show that they're not monitoring the main output. *Solo* is rather similar, while *solo in place* means that the stereo positioning of the selected source is maintained.

We can now produce a block diagram of a complete channel (without AFL, etc.), in Figure 21.3.

The reader will no doubt be cheered to know that we've done the hard part of sound desks; what follows is going to be quite easy!

21.3 Groups and busbars

Let's think about a typical and often frequent situation for the operator of a sound desk: we have a sizeable orchestra playing 'big band' music in the studio and it's being multi-miked. There could be perhaps 10 microphones on the violins, eight each on

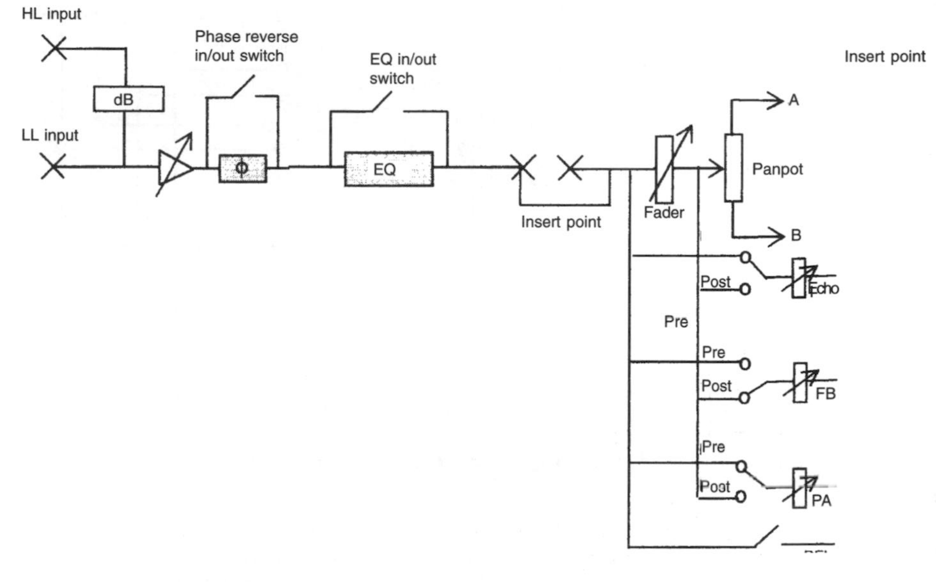

Figure 21.3 A nearly complete channel

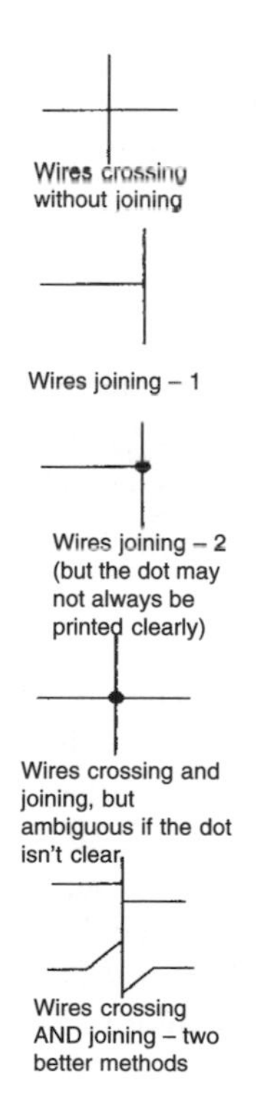

the brass and woodwind. There could be a drum kit with another eight microphones, and so on. The violin section is, at one stage, a little too loud and the operator has to reduce the settings of 10 separate microphone faders. Or it could be the drum kit which needs bringing up – eight faders. All of this is not easy. The answer is to make use of *groups* – all the microphone channels are grouped together so that one fader acts as a kind of master for the 10. Similarly the brass microphone channels can be grouped, and so on. This means having a set of faders called *group faders*. Of course, there has to be a way of switching channels to groups, and this is usually performed by a set of push buttons (our diagram, for simplicity, shows rotary switches). This function is called *channel routeing*. Next, we have to remember that there are going to be very many channels, all to be connected to the different groups. This involves *busbars*, which can be thought of as longish wires running through the desk each one providing a link between channels and groups. And while we're about it we might as well add that the feeds of PA, FB, echo, PFL, and so on, from each channel go on to their busbars. For the moment, though, to keep things simple, we'll start with the channel routeing. Figure 21.4 shows this. Note that we've shown only four groups, to avoid making the diagram look complicated. And, of course, there have to be two sets, one

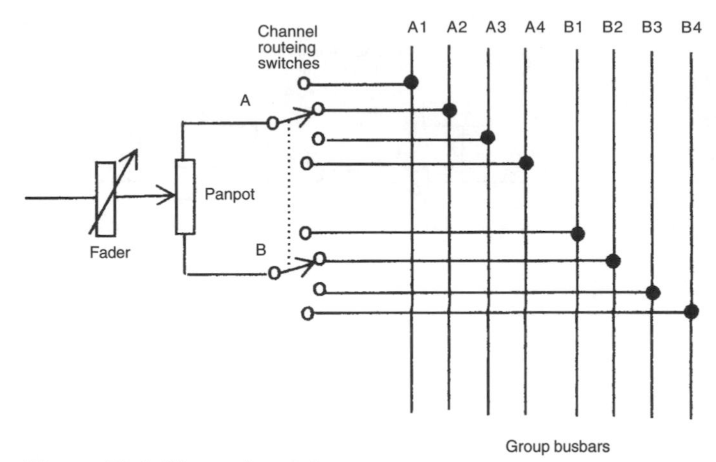

Figure 21.4 Channel routeing

for the stereo A signals and one set for the B signals. (The dotted line joining the A and B switches means that the switches are *ganged* together – one control button or knob operates both sections of the switch.)

Our margin diagrams show various ways of showing wires crossing and/or connecting. To emphasize things we've adopted here the convention of showing a dot where two wires join. In later diagrams we will only do that where it will make the situation clearer.

Our next step is to add the group faders, connected to the busbars, and now we'll add the busbars for PA, FB and PFL. But, to avoid a complicated diagram we'll show only a very little of the stereo B side. Also, on large desks, it's quite common for each group to have a range of facilities, insert points, EQ, and derivation of echo, PA and FB, similar to channels. The basic and simplified version is shown in Figure 21.5.

21.4 The output stages

We've now reached the point where we can bring things together to show the signals which leave the desk. The main thing, obviously, is for the group outputs to be linked and then go to an overriding fader. This may be known as the *main fader* or the *master fader*. (Don't forget that this, as with the group faders, must be a double one with tracks for the stereo A and B signals.) Figure 21.6 shows this. We've also added an outline of the monitoring which can take place here, but monitoring of

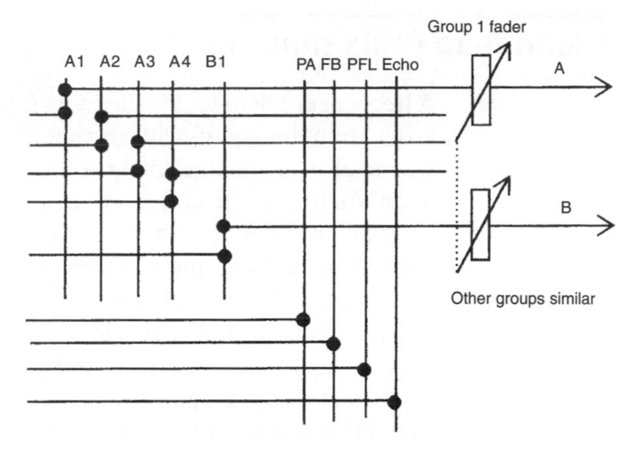

Figure 21.5 Group faders

individual channels or groups is also likely to be found. A further item in the diagram is a point where signals returning from an artificial reverberation device can be added to the main (*direct*) signal. So that the level of this can be controlled there is an *echo return fader*, but we should add that there can be other ways of bringing the 'echo' back into the main stream.

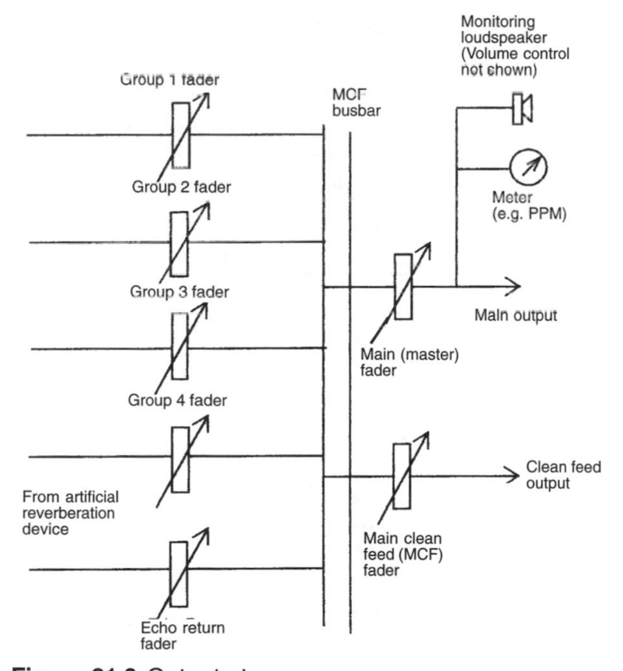

Figure 21.6 Output stages

21.5 Clean feed ('mix minus')

A frequent problem for broadcasting (radio or television) studios arises from linking up with another studio in a live situation. If we imagine a simple example of this we have a presenter in the main studio (we'll call it A) and he or she is linked to a contributor in a remote studio (B). Each obviously needs to hear what the other is saying and their conversation needs to go off to the transmitters. What may be very undesirable is for one, or possibly more than one, person to hear their own voice coming back to them. This is very much the case if the studios are any significant distance apart because there can then be disturbing time delays. The person in the B studio can find it almost impossible to talk if they hear their own voice delayed by a fraction of a second or more. In radio it's likely that both presenters would be wearing headphones; in television the other person's contribution could be heard in miniature (and inconspicuous) earpieces, which are going to be the equivalent of headphones, or there may be an out-of-vision loudspeaker. Either way a delay can be distressing. This is where *clean feed* (UK term; in the USA the equivalent is *mix minus*) becomes important.

A clean feed can be defined as *the output of the sound desk minus one or more of the sources going into it*. (Here we have to admit that the American term of 'mix-minus' is the more self-explanatory of the two!) Figure 21.7 can help to make this definition clear.

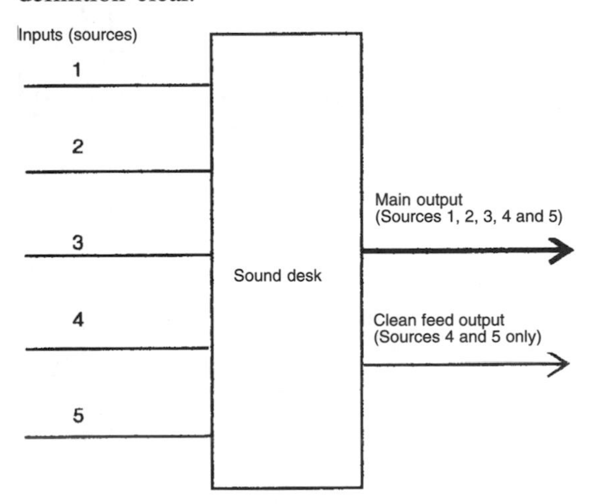

Figure 21.7 Illustrating clean feed

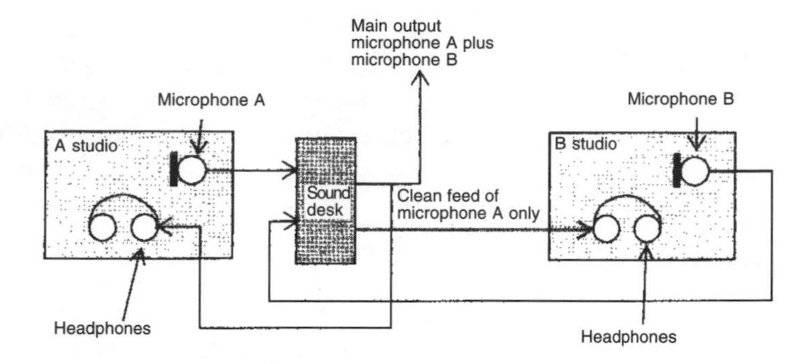

Figure 21.8 A fairly typical use of clean feed

There are five inputs to the desk – these could be from microphones, CD players – anything. The main desk output contains them all, assuming they're all faded up, of course! The clean feed output consists only of sources 4 and 5. All the others, 1, 2 and 3 in this example, are missing. In other circumstances a clean feed might consist of only one of the sources, or it might consist of everything except one source.

To illustrate the use of a clean feed in preventing the contributor in a remote studio from hearing his or her own voice, possibly delayed, look at Figure 21.8, which should now be self-explanatory.

Other uses of clean feeds include things like:

1. If a musical programme is being sent to another country as well as the UK, the other country may not want the UK presenter's comments and announcements. They might wish to insert their own language instead. The other country could then have a clean feed which could be all of the music but not the UK commentary.
2. Rather similar to item (1), in technical terms, is the case of a major sporting event of international interest. The clean feed to other broadcasters would be of crowd noise and other effects (often called *clean FX*).

So, in terms of the sound desk, sources to make up a clean feed are routed to a separate busbar and there will be a *main clean feed* (MCF) fader to control the overall level.

And that is more or less that! There can be an infinite number of variations on the desk we've outlined here, but this should be enough to start the reader off in the right direction. Also, *digital sound desks* are becoming more common and at lower prices. This is not the place to try to explain how they work but it is broadly true to say that, from the operator's point of view, the controls and their functions may not differ all that much from the analogue type we have been looking at.

21.6 Communications

There's one final and very important part of a sound desk and its associated apparatus that we must mention – methods of communicating with other people or areas. This has nothing directly to do with the main programme sound but because speech links are audio signal they have become part of the sound operational staff's problem.

In radio the communications are usually relatively uncomplicated, the simplest being *light signalling*. The standard is for *blue* lights both outside and inside the studio to indicate *rehearsal* conditions. This doesn't mean necessarily that there is a rehearsal in progress; it may only mean that the studio is powered (i.e. switched on).

Red lights indicate *transmission*, which can include recording. Red lights mean absolute silence, no going into the studio unless it is absolutely essential, and so on. Television studios use blue and red lights for the same purpose.

Green lights are used for *cueing* purposes. These are operated by small switches in the control room and can be used, for example, to tell a presenter to start talking.

Probably the next most important item of communication is *talkback*. In its simplest form in sound studios this consists of two microphones in the control room, one for the sound operator, built into the desk or on a flexible 'stalk' mounted on the desk, and one for the producer, again possibly on a stalk at the end of the sound desk or on a separate table. The talkback microphones are made live by a switch and this then allows the performers in the studio to hear instructions. There's an obvious problem here: if any of the studio microphones are faded up then the use of *talkback loudspeakers* is clearly undesirable! In such circumstances headphones must be worn. There's an automatic system which allows for all contingencies, depending on whether the studio is in transmission or rehearsal mode. In rehearsal mode the talkback loudspeakers can operate, but in transmission conditions the talkback loudspeakers are cut so headphones have to be used. If, though, no studio microphones are faded up then the loudspeaker talkback is operative.

Then there is *reverse talkback* (RTB) and this is simply a means of letting the performers in the studio talk to the people in the control room. Sometimes one of the ordinary studio microphones can be used for this when in rehearsal. Alternatively an RTB microphone may be built into, say, a musical director's desk.

Television communications are much more complicated because there are so many people and areas involved. There's still loudspeaker talkback, although this isn't often used except in the early stages of rigging the studio.

First there is *production talkback* or *open talkback*, which is unswitched and the microphone is therefore open all the time. The main user of this is the director who will be talking more-or-less constantly to the floor manager and most of the crew. *Sound talkback* allows communication with a boom operator, if there is one, who will also have his or her own reverse talkback. There'll also be talkback to, for example, the vision operators and video recording areas and *lighting talkback*, for the lighting director to use.

Finally we can mention *intercoms*, which are independent of talkback but allow communications between many areas, rather like an internal telephone system but more convenient as you don't have to hold a handset to your ear!

It wouldn't be overstating things greatly if we were to say that in a television studio effective communications along the lines we've mentioned are vital. Without them the whole system fails!

22

Safety

22.1 A little quiz

1. You are helping to rig some sound equipment in a television studio and there is a heavy loudspeaker which your supervisor asks you to put on to a trolley. Do you:

   (A) Demonstrate your keenness to help by struggling to lift it on your own.
   (B) Tell him or her to lift it themselves.
   (C) Explain that you need someone to help, and then, with that other person, put the loudspeaker on the trolley using correct lifting techniques.

2. A radio programme is to be recorded in a village hall. A cable needs to be taken past a doorway which is also a fire exit. You are given the job of laying that cable. Do you:

   (A) Hope that there isn't going to be a fire.
   (B) Fix a bit of paper nearby with the words 'Beware of the cable'.
   (C) Either take the cable over the doorway and fix it securely so that it can't fall, or if that isn't possible, lay it flat on the floor and cover it with a piece of rug or a doormat so that no one can trip over it.

3. An item of mains-operated equipment has failed and the indications are that the fuse has blown. Which course of action do you adopt?

   (A) Remove the cover to see if you can spot what's wrong.
   (B) Remove the fuse and check it.
   (C) Switch off and unplug the equipment from the mains. Then check the fuse.

4. You are on a gantry some 5 m above the floor of a television studio and realize that the spanner you need to undo a nut is on the studio floor below you. Do you:

   (A) Go away and hope that someone else will do the job.
   (B) Shout to someone below to throw you the spanner.
   (C) Go down to fetch the spanner.

5. Someone asks you where the studio fire extinguishers are. You don't know. Do you:

   (A) Shrug your shoulders and say it's nothing to do with you.
   (B) Pretend you haven't heard and walk away.
   (C) Admit you don't know but then find out as soon as possible.

6. A notice appears in the crew room announcing that there will be instruction in resuscitation next Thursday morning at 09:30 hours and all staff who can be free should attend. You have no duties scheduled for that time. Do you:

   (A) See this as a good opportunity to have an extra hour or two in bed and not come in until later.
   (B) Decide that since it's very rare for anyone in the organization to be electrocuted it will be a waste of time and instead you can use that morning to work out your expenses.
   (C) Realize that the next person to get a severe electrical shock *could* be *you* and it's in your interests if the people you work with know how to deal with this. You therefore have a moral duty, as well as possibly a legal responsibility, to know how to resuscitate a person.

The right answer in every case is clearly C and without doubt you will have scored the maximum number of points. The quiz was slightly facetious but nevertheless it goes some way towards showing that, to a large extent, good safety is a matter of common sense.

It's also vital to realize that taking a little longer to do something may well, in the long run, be easier and cheaper! Take Question (4), for example. To get someone to throw the spanner to you might save a minute or two compared with going down to the studio floor, picking up the spanner and going back on to the gantry. *But* suppose someone threw the spanner and you failed to catch it! Think of the possibilities:

1. The spanner might hit and damage a piece of equipment. This could be expensive (very!) and could delay the start of rehearsals until a replacement was found.
2. The spanner might hit someone. Imagine the delay while first aid was rendered, and if the injury was severe enough a replacement person would have to be found while the injured one was taken off to the surgery, or worse still, to hospital.
3. Your own position, and that of the spanner thrower, would be dubious. You'd probably both have a very unpleasant interview with a supervisor and you might be lucky if the matter ended there. It would have been much easier to have fetched the spanner yourself!

There are a few extra points to note and we'll deal with these in the rest of this chapter, but good safety practice is, as much as anything, a state of mind.

SAFE THINKING = SAFE WORKING

22.2 Electrical safety

Electricity can kill! Some good basic rules are:

1. If in doubt about equipment switch it off and remove the plug from the supply.
2. Always make sure that plugs are wired correctly with the *live* (brown) wire going to the correct terminal (the one with the fuse), the *neutral* (blue) going to the terminal marked N and the *earth* (yellow/green) going to the terminal near the apex of the triangle of pins. The diagram in the margin will remind you.

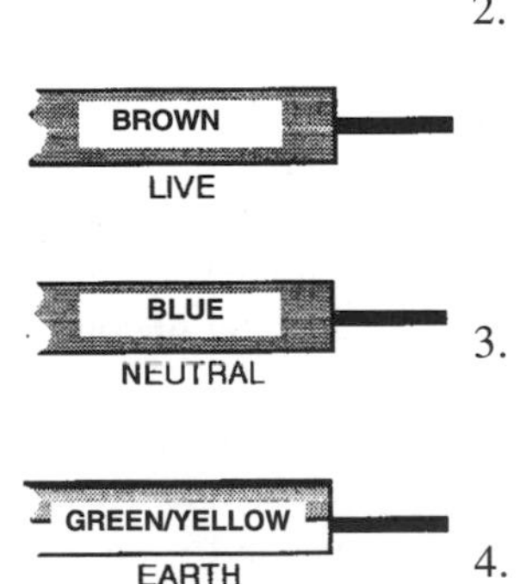

3. Check all plugs and sockets periodically. Reject any where bare conductor is exposed, where the fixing screws are not tight and where there are any signs of damage. See that the cable grips are doing their job.
4. Always make sure that fuses are of the appropriate rating.

We saw earlier in this book how to do a very simple calculation to find the correct rating.

5. Check cables for signs of wear or damage.
6. Where mains-operated equipment has to be used out of doors, take extra care. Isolating transformers should be used (only one piece of equipment to each transformer), or perhaps devices of the power breaker type can be used as an alternative.

The list is almost endless but the few we have given here, if followed, will help.

Resuscitation

Everyone who works in a technical area should know how to apply life-saving methods to a person suffering from an electric shock. Periodical training or refresher sessions should be held with qualified instructors. And, besides knowing how to treat a victim it's essential to know how to protect yourself first – for example by pulling the person away from live cables, using part of their clothing, so that you don't get an electric shock yourself. It all sounds so obvious but at times of stress it's all too easy to forget the obvious precautions.

22.3 Lifting and handling

This means knowing how to avoid injury to yourself or anyone else when dealing with heavy or cumbersome items. A tremendous number of people suffer from back trouble, which can be very painful and unpleasant even if it only lasts a few days. In more serious cases the trouble can last for months, or even be permanent. And incorrect lifting is a very common cause.

The human spine is perhaps not too well designed for modern life styles. It's all right providing that it can be kept straight, or very nearly so. Putting a strain on it when it's bent, as can easily happen when we are lifting things, can cause damage – maybe permanently.

The answer is never to lift heavy objects on one's own, and when something *has* to be lifted, whether with help or not, always keep the back straight. A good rule is:

Bend the knees and not the back.

Studios and other technical areas should have posters showing this in more detail and proper instruction should be provided at intervals. If it isn't then ask for it!

22.4 Scaffolding, ladders and rigging generally

All these are potentially hazardous! Scaffolding is commonly used in outside broadcast work, for example to provide elevated platforms for cameras. These structures should be erected by experts – specialized contractors, for example – but it's as well for everyone to know that there should be safety boards round the base of the platform to prevent anyone's feet from slipping over the edge and also to ensure that tools cannot roll off and hit someone below.

Ladders must be firmly lashed at the top so that they can't slip and they should be at an angle of about 75° (a slope of 1 in 4). This is not so steep that there is a risk of the ladder falling backwards but at the same time it is unlikely to slide down the wall. Frictional forces are optimum at this sort of angle.

Then, great care is needed in rigging equipment which is to be slung up in the air. Safety chains, which will restrain the item if the main fixing clamps fail, are essential. When microphones are suspended over an audience the microphone cable should be used as the equivalent of a safety chain to support the microphone if its clips give way. (Some authorities may not allow microphones to be hung over an audience.)

As we've suggested in the quiz, cables going past doorways should ideally be taken over the door but if that's impossible then they must be fixed to the floor using special adhesive tape (*gaffer tape*) and covered with mats or carpet so that it's impossible for anyone to trip over them.

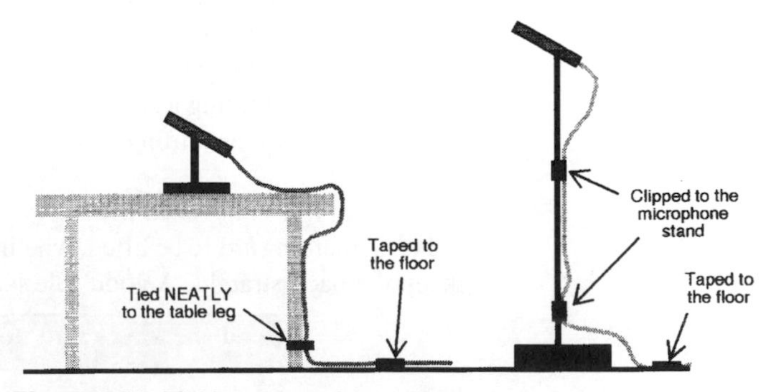

Figure 22.1 Making microphone cables safe

Microphones, whether on table stands or full length floor stands must have the cables tied or clipped as far down as possible so that it's not going to be easy for anyone to catch their feet in them. And then the cable should be taped to the floor. (See Figure 22.1.)

22.5 Ten important extras to look out for

1. Spilt tea or coffee on a hard floor is slippery and therefore dangerous.
2. Razor blades (for editing analogue tape) must be disposed of safely – they must *never* be left lying around. Special containers for used blades should be near at hand.
3. Make sure you know where the nearest first aid box is and what it should contain. If it's incomplete inform your supervisor.
4. Know how to fill in accident report forms.
5. Report *any* accident, no matter how trivial.
6. Don't forget that high noise levels can cause permanent hearing damage.
7. Fire doors must never be wedged open (least of all with a fire extinguisher, which seems to be a common fault!)
8. Know where the fire alarms are and where the assembly points are if a fire occurs.
9. Television studios have *fire lanes* – passageways behind the scenery to allow easy access to fire exits. These must never be obstructed.
10. Know how to contact the necessary people in an emergency – the first aiders, the fire wardens, and so on.

23

Further reading

There are large numbers of books on the subject of sound in radio, television and recording studios. Many are good, although the reader shouldn't assume that because something is in print it is necessarily reliable in its content! (This applies to magazine articles as well.) And unfortunately much of the really worthwhile material isn't always very easy to read. The short list below gives some books which aren't difficult to read and at the same time contain useful and reliable material.

Alkin, Glyn, *Sound Techniques for Video and TV*, 2nd edn, Focal Press, 1989, Oxford.

Alkin, Glyn, *Sound Recording and Reproduction*, 3rd edn, Focal Press, 1996, Oxford.

Bermingham, A. et al. *The Video Studio*, 3rd edn, Focal Press, 1994, Oxford.

Lyver, Des, *Basics of Video Sound*, Focal Press, 1995, Oxford.

Nisbett, A. *The Use of Microphones*, 4th edn, Focal Press, 1994, Oxford.

Nisbett, A. *The Sound Studio*, 6th edn, Focal Press, 1995, Oxford.

Rumsey, F. & McCormick, T. *Sound and Recording: An Introduction*, 3rd edn, Focal Press, 1997, Oxford.

Talbot-Smith, M. *Broadcast Sound Technology*, 2nd edn, Focal Press, 1995, Oxford.

Talbot-Smith, M. *The Sound Engineer's Pocket Book*, Focal Press, 1995, Oxford.

Watkinson, John, *An Introduction to Digital Audio*, Focal Press, 1994, Oxford.

24

Glossary

(Words or phrases in italics are explained elsewhere in this Glossary.)

A

Absorption coefficient. The fraction of sound which is absorbed by a material which it strikes.

a.c. Alternating current. Used for currents or voltages which reverse direction at a regular rate. The rate at which the direction changes is the *frequency*.

ADC. (See *Analogue to digital converter.*)

AES. Audio Engineering Society.

a.m. (See *Amplitude modulation.*)

Ampere. Unit of electric current. It depends on the number of electrons flowing per second.

Amplitude. The 'height' of a waveform above its zero line, measured in the appropriate units.

Amplitude modulation. A way of transmitting audio and other signals by varying the amplitude of a high frequency 'carrier'.

Analogue. Analogue audio signals are voltages which are proportional to the pressures in the original sound wave.

Analogue to digital converter. Device which accepts analogue signals and converts them into digital ones. The reverse process is carried out by a *digital to analogue converter, (DAC)*.

Antinode. Region in a resonant vibrating system where there is a maximum vibration. (See *Node.*)

Attenuator. Circuit, usually of *resistors*, intended to reduce signal levels, typically to avoid overloading a piece of equipment.

Azimuth. In analogue tape recorders the azimuth of the heads is correct if their gaps are exactly at right angles to the direction of the tape.

B

Baffle. Usually a large wooden board with a hole in it for a loudspeaker *drive unit.*

Balance (in sound). The relative loudnesses of different parts of, for example, a band or orchestra.

Balanced (wiring). Cables, etc., in which pairs of conductors are electrically symmetrical, thus reducing the effects of induced interference.

Bass. Imprecise term for the low frequency end of the audio range.

Bass reflex loudspeaker. (See *Vented enclosure.*)

Bass tip-up. Effect shown by *pressure gradient* microphones in which there is a marked increase in the bass output of the microphone when sound sources are close to the microphone. Sometimes called 'proximity effect'.

Battery. A number of electrical cells joined together.

Bi-amp system. Arrangement for feeding multiple unit loudspeakers with two separate amplifiers.

Bi-directional. (See *Figure-of-eight.*)

Binary arithmetic. System of arithmetic using 2 as a base instead of 10. Binary numbers are thus represented only by 1s and 0s.

Bit. Binary digit. A 1 or a 0 in *binary arithmetic.*

Bit rate. The number of *bits* being recorded, transmitted, etc., each second.

Busbar. Basically a wire or other conductor in, for example, a sound desk to which may be connected a number of signals.

Byte. Set of eight *bits*.

C

Capacitor. Electrical component which can store electrical charges. These are minute compared with the storage capabilities of a battery. The electrical 'size' of a capacitor is called its 'capacitance'. The term 'condenser' is an obsolete alternative.

Capsule. *Capacitor/diaphragm* assembly in an *electrostatic* microphone.

Cardioid. (1) *Polar diagram* which is heart-shaped. **(2)** Microphone with a cardioid *polar diagram.*

Channel. (1) One side (left or right) of a *stereo* system. **(2)** Signal path in a sound desk handling the output of one source, such as a microphone.

Chip. (See *Integrated circuit.*)

Clean feed. The output of a sound desk but without one or more of the sources fed into the desk. Also known, especially in the USA, as 'mix minus'.

Coincident pair. A pair of closely-spaced microphones used in *stereo* operations.

Colouration. Generally unwanted minor but audible alterations to the quality of a sound, e.g. by small resonances in the cabinet of a loudspeaker.

Column loudspeaker. (See *Line source loudspeaker*.)

Companding. Term used for compressing and later expanding the dynamic range of a signal.

Complex wave. A repeating waveform which is not a sine wave.

Compressor. Circuit which provides a degree of electronic control of the level of an audio signal.

Condenser microphone. (See *Electrostatic microphone*.)

Coulomb. Unit of charge of electricity.

Crossover unit. Circuit in a multiple unit loudspeaker which sends different bands of audio frequencies to the appropriate *drive units*.

Crosstalk. The unwanted transfer of signals from, for example, one track to another on a tape machine.

D

dB. Abbreviation for *decibel.*

dBA. Form of *decibel* which approximates to loudness.

DAC. (See *Analogue to digital converter.*)

DAT. Digital audio tape recording.

Decibel. Unit of comparison of two powers, two voltages, two sound pressures, etc.

DI box. (See *Direct injection box.*)

Diaphragm. Thin membrane of plastic or metal in, for example, a microphone. The diaphragm moves when sound waves strike it and this movement causes a small voltage to be generated in a *transducer.*

Diffraction. The bending of waves round an obstacle, given the right conditions.

Diffraction grating. Transparent screen, (or reflecting surface) ruled with precisely spaced fine lines which split a light beam into a coloured spectrum. If the light is very pure, as from a laser, it is split into beams.

Diffusers. Irregular surfaces used in studios etc. to reduce the effect of *standing waves.*

Digital. Digital signals are in the form of *pulses* of voltage, all of the same amplitude but bearing no resemblance to the original. The relationship between digital signals and their original analogue form is an arithmetical one.

Direct current (d.c.). A current or voltage which flows in one direction only.

Direct injection box. Unit used to connect an electrical musical instrument directly to a sound desk instead of having a microphone.

Direct sound. Sound which travels without reflection from source to microphone. (See also *Indirect sound.*)

Directivity pattern. (See *Polar diagram.*)

Dolby. A widely used range of systems, primarily used for reducing noise in analogue tape recordings.

Drive unit. The coil, magnet and cone assembly of a loudspeaker.

Drop out. In magnetic tape a region where the tape coating is thin or missing so that recorded signals in this area are low in level or even absent.

E

EBU. European Broadcast Union. Organization formed from broadcasters in Europe and responsible, amongst other things, for setting technical standards.

Electron. A fundamental particle inside an atom. It has a negative charge and is considered to orbit the nucleus. An electric current consists of a flow of electrons.

Electret. Type of *electrostatic microphone* using a permanently charged *capacitor* in the *capsule.*

Electromagnetic waves (e.m. waves). Waves which are partly electrical and partly magnetic in their character. They travel through space at 300 000 km/s. Radio waves and light waves are e.m. waves.

Electromotive force (e.m.f.). The voltage at the terminals of a battery or other generator when no current is being drawn from it.

Electrostatic microphone. Microphone which operates by sound pressures affecting one plate of a *capacitor*.

Enclosure. Word often used to mean the cabinet of a loudspeaker.

EQ. Abbreviation for *equalizer*.

Equalizer. Circuit which can emphasize or reduce specific frequency ranges. Abbreviation is EQ.

F

Fader. Control, usually in the form of a slider, for altering the level of an audio signal.

Farad (F). Unit of capacitance. (See *Capacitor*.)

FB (See *Foldback*.)

Figure-of-eight. Term for the *polar diagram* of a microphone which is sensitive to sounds from the front and the rear but not from the sides. Sometimes called 'bi-directional'.

Flux density. (See *Magnetic flux density*.)

f.m. (See *Frequency modulation*.)

Foldback. System for allowing musicians, actors, etc., to hear feeds of selected sound sources via headphones or loudspeakers.

Frequency (*f*). The number of times a second a waveshape is repeated. The unit is the Hertz (Hz). (See *kilo, mega*.)

Frequency modulation (f.m.). A method of transmitting audio and other signals by varying the frequency of a high frequency 'carrier'. (Compare *Amplitude modulation*.)

Frequency response. Method of indicating one aspect of the performance of a microphone or loudspeaker by showing how its output varies over the required frequency range.

Fundamental. The lowest frequency in a complex wave.

G

Gain. The change (increase or decrease) in the level of a signal expressed in decibels.

Groups. In a sound desk the outputs of two or more channels can be brought together with a group fader to act as a sub-master control.

H

Harmonic. A component of a *complex wave*. Harmonics have frequencies which are exact multiples of the *fundamental* frequency.

Helmholtz resonator. Acoustic resonator consisting of an air-filled vessel with a small open neck.

Horn unit. Type of *drive unit* used in some loudspeakers.

Hypercardioid. *Polar diagram* intermediate between *figure-of-eight* and *cardioid*.

I

Integrated circuit (IC). A collection of possibly very large numbers of *transistor*-type devices in one small unit. Often called a 'chip'.

Indirect sound. Sound which reaches a microphone having undergone one or more reflections. (See also *Direct sound.*)

Inductor. Component, almost invariably a coil of wire, often wound on an iron core, and intended to provide *reactance*.

Inductance. The electrical 'size' of an *inductor* is called its inductance.

Insert point. Point in, for example a sound desk, where additional equipment may be inserted in the signal path.

Intensity. With waves, the power falling or passing through an area. Basic unit is watts/square metre. It is *not* the same thing as loudness.

Inverse square law. Law relating the way in which the *pressure* or *intensity* of a wave decreases with distance from the source.

Integrated services digital network (ISDN). A method of using standard telephone systems for carrying digital audio signals.

J

Jack. Plug used for temporary connection of audio equipment.

K

Kilo (k). Prefix meaning a thousand.

L

Laser. Device for producing light consisting of one wavelength only.

Light emitting diode (LED). Small electronic device which can produce light when a suitable current flows through it.

Limiter. Device similar to a *compressor* but with a fixed upper limit for the signal level.

Line-up. Process of ensuring that different items of equipment in a chain are working at the same, or at least compatible, levels.

Line source loudspeaker. Type of loudspeaker with directional properties. Also known as 'column loudspeaker'.

Longitudinal wave. Type of wave in which the particles of the medium vibrate in the same direction as the wave. Sound waves are of this type.

Loudness. A subjective assessment or impression of the 'strength' of a sound.

M

Magnetic field. The region around a magnetized object where its effects can be detected.

Magnetic flux density. Roughly speaking, the strength of a magnetic field at a particular point.

Mega (M). Prefix meaning a million.

micro (m). Prefix meaning one millionth.

MIDI. Musical instrument digital interface. A standard used for linking electronic musical instruments together and/or with computers.

milli (m). Prefix meaning one thousandth.

Mix minus. (See *Clean feed.*)

Modulation. A process in which the form of a signal is radically changed for the purposes of, say, transmission or recording. (See *Amplitude modulation* and *Frequency modulation.*)

Moving coil microphone. A microphone in which the transducer consists of a coil fixed to the diaphragm, the coil being in a magnetic field.

N

Newton (N). Unit of force. It is that force which would give a mass of one kilogram an acceleration of one second each second.

NICAM. Method of reducing the *bit-rate* in a *digital* system.

NICAM 728. System used to transmit *stereo* sound with television pictures.

Node. A region in a resonant vibrating system where there is a minimum of vibration. This may depend on what is being studied. For example in high frequency electrical transmission it is possible for a voltage node to occur at a current maximum (antinode).

Nucleus. The core of an atom around which it is assumed that electrons move in orbits.

O

Octave. An interval of musical pitch. Two notes an octave apart are accepted by the ear as having something in common. A change of one octave is equivalent to a doubling or halving of the frequency.

Ohm (Ω). Unit of electrical resistance.

Ohm's law. Relates volts, amperes and ohms by $V = IR$, $I = V/R$ or $R = V/I$.

Omnidirectional. A microphone, for example, has an omnidirectional response if it is equally sensitive (or nearly so) to sounds from all directions. Often abbreviated to 'omni'.

P

PA. Public address.

PCM. Pulse code modulation. Really another term for digital technology.

PPM. (See *Peak programme meter.*)

Pan pot. Control used for positioning mono sound images in a *stereo* 'scene'.

Parity. A basic method of detecting errors in a *digital* signal.

PASC. Precision adaptive sub-band coding. A method of reducing the *bit rate* by recording only the part of the audio signal which will be detected by the normal ear.

Pascal. Unit of *pressure*, equal to 1 newton/square metre.

Peak programme meter (PPM). Device for indicating the peak levels of an audio signal.

Peak-to-peak value. The measurement of a waveform from peak to trough. With a sine wave or other symmetrical waveform it is twice the *amplitude*.

Perspective, sound. (See *Sound perspective.*)

PFL. Pre-fade listen. The ability to monitor sound sources before they are faded up. Similar to prehear.

PH, Pre-hear. (See *PFL.*)

Phantom power. Method of supplying electrical power to microphones along normal three-conductor microphone cable.

Phase angle. If two sine waves are of the same frequency but are not 'in step' the phase angle gives the amount by which they are 'out of step'. A complete cycle is regarded as covering 360°. Thus if two waves are displaced by a quarter of a cycle they are said to out of phase by, or have a phase difference of, 90°.

Phase difference. (See *Phase angle.*)

Polar diagram. Diagram showing how the response of a microphone or loudspeaker changes with direction.

Power. The rate of doing work. The unit is the watt.

Pressure. The force acting on a unit area. The unit in sound is the *pascal* (Pa).

Pressure gradient. This usually refers to a type of microphone in which the *diaphragm* movements depend on the sound pressure difference between its two sides.

Pressure zone microphone (PZM). This consists of a small microphone unit placed very close to, or in the face of, a relatively large area surface.

Primary cell. Device which produces an *e.m.f.* by chemical action. These cells cannot be recharged when they are 'spent' – meaning that one or other of its chemical constituents have been used up. So-called 'dry cells' are primary cells.

Proximity effect. (See *Bass tip-up.*)

Pulse. An electrical signal of fixed amplitude and duration and used to represent a *binary digit.*

PZM. (See *Pressure zone microphone.*)

R

RAM. Random access memory. Usually a *chip* which can store digital signals but only as long as it is powered.

R-DAT. The same thing as DAT.

Reactance. The equivalent of *resistance* in a.c. circuits provided by *inductors* and *capacitors*. Reactance normally varies with the frequency of the a.c.

Rectifier. Component which allows electricity to flow through it in one direction only.

Resistance. The property of restricting the flow of an electric current.

Resistor. Component designed to provide a specified amount of *resistance.*

Resonance. The condition in which a mechanical structure, electrical circuit, acoustic system, etc., is set into its own natural frequency of oscillation. The amplitude of the oscillation at resonance is likely to be much greater than the amplitude of the vibration causing it.

Resuscitation. Reviving a person suffering from electric shock, drowning, etc.

Reverberation. The dying away of sound in a room, hall, studio, etc.

Reverberation time (r.t. or T_{60}). The time in seconds taken for the reverberant sound level to decay through 60 *decibels*.

Root-mean-square (r.m.s.). The r.m.s. value of a current or voltage is that which would have the same heating effect as a d.c. current or voltage of that value. The a.c. mains in the UK has an r.m.s. voltage of 230 although the *amplitude* is about 325 V.

S

Sabine. Unit of sound absorption. It is equivalent to one square metre of 100% absorber.

Secondary cell. Electrical device which, unlike a primary cell, can be recharged. Car batteries and NICAD cells are good examples.

Semitone. A specific change in the musical pitch of a sound. It corresponds fairly closely to a frequency change of 6%. There are 12 equal semitones to an *octave*.

Sensor. Device which can detect, for example, light.

Short circuit. A path of very low electrical resistance across a pair of wires or the terminals of a generator, effectively preventing a current from flowing anywhere else.

Sidebands. Additional frequencies produced in *modulation* processes.

Sine wave. A very pure waveform representing only one frequency.

SMPTE. Society of Motion and Television Picture Engineers. An American organization which has given its name to certain standards. (See also *EBU*.)

Sound perspective. The apparent distance, judged aurally, of a sound source.

Standing wave. An acoustic *resonance* between two flat parallel surfaces. It is called this because the *nodes* and *antinodes* have fixed positions.

Starting transients. Frequencies at the start of a sound which die away quickly and which are not arithmetically related to the *fundamental*. They are very important in helping the brain to identify musical instruments, etc.

Star-quad. Type of microphone cable which has a high immunity to the effects of external interference.

Stereo. Short for ‘stereophony’. System using two audio channels and two spaced loudspeakers to give a correctly-sited listener an impression of width of the sound images.

Sub-woofer. Loudspeaker unit intended to produce extra low frequencies.

T

Talkback. Method of allowing communication between production staff and others.

Thermionic valve. Electronic device capable amongst other things of amplifying electrical signals. To a very large extent valves have been replaced by *transistors*.

Threshold of audibility (or hearing). The lower limit of sound pressure which can be detected by the normal ear.

Time code. Digital signals which can be recorded to give very accurate information about timing.

Timbre (pronounced ‘tarmbre’). The tonal quality of a sound as judged by the ear.

Top. Imprecise term for the upper end of the audio frequency range.

Transducer. Components which convert electrical energy into mechanical movement, or the other way round.

Transformer. Electrical device consisting in its simplest form of two coils, usually wound on an iron core. An a.c. voltage fed into one coil appears at the other coil with a different voltage which depends on the numbers of turns in the two coils – the 'turns ratio'.

Transients. (See *Starting transients.*)

Transistor. Small electronic component capable of amplification, amongst other things.

Transverse wave. Type of wave in which the particles of the medium vibrate at right angles to the direction of the wave. Waves in water are transverse waves.

Treble. (See *Top.*)

Turns ratio. (See *Transformer.*)

Tweeter. Loudspeaker *drive unit* designed to work at the higher audio frequencies.

V

VCA. (See *Voltage controlled amplifier.*)

Valve. (See *Thermionic valve.*)

Vented enclosure. Loudspeaker using a *Helmholtz resonator* to improve its performance at low frequencies. Sometimes called 'bass reflex'.

Volt. Unit of electrical pressure.

Voltage controlled amplifier (VCA). Amplifier, usually in the form of a chip, whose gain is controlled by a voltage applied to it.

VU meter. Device for indicating programme signal levels. (VU = Volume unit.)

W

Wavelength. The distance between corresponding points in a wave. Denoted by the Greek letter λ (lambda). Wavelength is related to frequency and wave velocity by the formula $c = f\lambda$.

Watt. Unit of electrical power.

Woofer. Term sometimes used for a loudspeaker *drive unit* designed to cover the low frequency end of the audio range.

Z

Zero level. Widely accepted voltage used as a reference with which audio signals can be compared using decibels. Equal to 0.775 V.

Index